Teacher Edition

Eureka Math
Grade 8
Module 6

Special thanks go to the Gordon A. Cain Center and to the Department of Mathematics at Louisiana State University for their support in the development of *Eureka Math*.

For a free *Eureka Math* Teacher Resource Pack, Parent Tip Sheets, and more please visit www.Eureka.tools

Published by Great Minds®.

Copyright © 2018 Great Minds®. No part of this work may be reproduced, sold, or commercialized, in whole or in part, without written permission from Great Minds®. Noncommercial use is licensed pursuant to a Creative Commons Attribution-NonCommercial-ShareAlike 4.0 license; for more information, go to http://greatminds.org/copyright. *Great Minds* and *Eureka Math* are registered trademarks of Great Minds®.

Printed in the U.S.A.

This book may be purchased from the publisher at eureka-math.org.

10 9 8 7 6 5 4 3 2 1

ISBN 978-1-64054-366-9

G8-M6-UTE-1.3.0-05.2018

***Eureka Math: A Story of Ratios* Contributors**

Michael Allwood, Curriculum Writer
Tiah Alphonso, Program Manager—Curriculum Production
Catriona Anderson, Program Manager—Implementation Support
Beau Bailey, Curriculum Writer
Scott Baldridge, Lead Mathematician and Lead Curriculum Writer
Bonnie Bergstresser, Math Auditor
Gail Burrill, Curriculum Writer
Beth Chance, Statistician
Joanne Choi, Curriculum Writer
Jill Diniz, Program Director
Lori Fanning, Curriculum Writer
Ellen Fort, Math Auditor
Kathy Fritz, Curriculum Writer
Glenn Gebhard, Curriculum Writer
Krysta Gibbs, Curriculum Writer
Winnie Gilbert, Lead Writer / Editor, Grade 8
Pam Goodner, Math Auditor
Debby Grawn, Curriculum Writer
Bonnie Hart, Curriculum Writer
Stefanie Hassan, Lead Writer / Editor, Grade 8
Sherri Hernandez, Math Auditor
Bob Hollister, Math Auditor
Patrick Hopfensperger, Curriculum Writer
Sunil Koswatta, Mathematician, Grade 8
Brian Kotz, Curriculum Writer
Henry Kranendonk, Lead Writer / Editor, Statistics
Connie Laughlin, Math Auditor
Jennifer Loftin, Program Manager—Professional Development
Nell McAnelly, Project Director
Ben McCarty, Mathematician
Stacie McClintock, Document Production Manager
Saki Milton, Curriculum Writer
Pia Mohsen, Curriculum Writer
Jerry Moreno, Statistician
Ann Netter, Lead Writer / Editor, Grades 6–7
Sarah Oyler, Document Coordinator
Roxy Peck, Statistician, Lead Writer / Editor, Statistics
Terrie Poehl, Math Auditor
Kristen Riedel, Math Audit Team Lead
Spencer Roby, Math Auditor
Kathleen Scholand, Math Auditor
Erika Silva, Lead Writer / Editor, Grade 6–7
Robyn Sorenson, Math Auditor
Hester Sutton, Advisor / Reviewer Grades 6–7
Shannon Vinson, Lead Writer / Editor, Statistics
Allison Witcraft, Math Auditor

Julie Wortmann, Lead Writer / Editor, Grade 7
David Wright, Mathematician, Lead Writer / Editor, Grades 6–7

Board of Trustees

Lynne Munson, President and Executive Director of Great Minds
Nell McAnelly, Chairman, Co-Director Emeritus of the Gordon A. Cain Center for STEM Literacy at Louisiana State University
William Kelly, Treasurer, Co-Founder and CEO at ReelDx
Jason Griffiths, Secretary, Director of Programs at the National Academy of Advanced Teacher Education
Pascal Forgione, Former Executive Director of the Center on K-12 Assessment and Performance Management at ETS
Lorraine Griffith, Title I Reading Specialist at West Buncombe Elementary School in Asheville, North Carolina
Bill Honig, President of the Consortium on Reading Excellence (CORE)
Richard Kessler, Executive Dean of Mannes College the New School for Music
Chi Kim, Former Superintendent, Ross School District
Karen LeFever, Executive Vice President and Chief Development Officer at ChanceLight Behavioral Health and Education
Maria Neira, Former Vice President, New York State United Teachers

This page intentionally left blank

A STORY OF RATIOS

Mathematics Curriculum

GRADE 8 • MODULE 6

Table of Contents[1]

Linear Functions

Module Overview 2

Topic A: Linear Functions 7

    Lesson 1: Modeling Linear Relationships 8

    Lesson 2: Interpreting Rate of Change and Initial Value 20

    Lesson 3: Representations of a Line 28

    Lessons 4–5: Increasing and Decreasing Functions 39

Topic B: Bivariate Numerical Data 67

    Lesson 6: Scatter Plots 68

    Lesson 7: Patterns in Scatter Plots 79

    Lesson 8: Informally Fitting a Line 95

    Lesson 9: Determining the Equation of a Line Fit to Data 107

Mid-Module Assessment and Rubric 119
Topics A through B (assessment 1 day, return 1 day, remediation or further applications 1 day)

Topic C: Linear and Nonlinear Models 129

    Lesson 10: Linear Models 130

    Lesson 11: Using Linear Models in a Data Context 142

    Lesson 12: Nonlinear Models in a Data Context (Optional) 153

Topic D: Bivariate Categorical Data 167

    Lesson 13: Summarizing Bivariate Categorical Data in a Two-Way Table 168

    Lesson 14: Association Between Categorical Variables 180

End-of-Module Assessment and Rubric 190
Topics A through D (assessment 1 day, return 1 day, remediation or further applications 1 day)

---

[1]Each lesson is ONE day, and ONE day is considered a 45-minute period.

Module 6:   Linear Functions

©2018 Great Minds®. eureka-math.org

Grade 8 • Module 6
Linear Functions

OVERVIEW

In Grades 6 and 7, students worked with data involving a single variable. This module introduces students to bivariate data. Students are introduced to a function as a rule that assigns exactly one value to each input. In this module, students use their understanding of functions to model the relationships of bivariate data. This module is important in setting a foundation for students' work in Algebra I.

Topic A examines the relationship between two variables using linear functions. Linear functions are connected to a context using the initial value and slope as a rate of change to interpret the context. Students represent linear functions by using tables and graphs and by specifying rate of change and initial value. Slope is also interpreted as an indication of whether the function is increasing or decreasing and as an indication of the steepness of the graph of the linear function. Nonlinear functions are explored by examining nonlinear graphs and verbal descriptions of nonlinear behavior.

In Topic B, students use linear functions to model the relationship between two quantitative variables as students move to the domain of statistics and probability. Students make scatter plots based on data. They also examine the patterns of their scatter plots or given scatter plots. Students assess the fit of a linear model by judging the closeness of the data points to the line.

In Topic C, students use linear and nonlinear models to answer questions in context. They interpret the rate of change and the initial value in context. They use the equation of a linear function and its graph to make predictions. Students also examine graphs of nonlinear functions and use nonlinear functions to model relationships that are nonlinear. Students gain experience with mathematical modeling.

In Topic D, students examine bivariate categorical data by using two-way tables to determine relative frequencies. They use the relative frequencies calculated from tables to informally assess possible associations between two categorical variables.

Focus Standards

Use functions to model relationships between quantities.

- Construct a function to model a linear relationship between two quantities. Determine the rate of change and initial value of the function from a description of a relationship or from two (x,y) values, including reading these from a table or from a graph. Interpret the rate of change and initial value of a linear function in terms of the situation it models, and in terms of its graph or a table of values.

- Describe qualitatively the functional relationship between two quantities by analyzing a graph (e.g., where the function is increasing or decreasing, linear or nonlinear). Sketch a graph that exhibits the qualitative features of a function that has been described verbally.

Investigate patterns of association in bivariate data.

- Construct and interpret scatter plots for bivariate measurement data to investigate patterns of association between two quantities. Describe patterns such as clustering, outliers, positive or negative association, linear association, and nonlinear association.

- Know that straight lines are widely used to model relationships between two quantitative variables. For scatter plots that suggest a linear association, informally fit a straight line, and informally assess the model fit by judging the closeness of the data points to the line.

- Use the equation of a linear model to solve problems in the context of bivariate measurement data, interpreting the slope and intercept. *For example, in a linear model for a biology experiment, interpret a slope of* 1.5 cm/hr *as meaning that an additional hour of sunlight each day is associated with an additional* 1.5 cm *in mature plant height.*

- Understand that patterns of association can also be seen in bivariate categorical data by displaying frequencies and relative frequencies in a two-way table. Construct and interpret a two-way table summarizing data on two categorical variables collected from the same subjects. Use relative frequencies calculated for rows or columns to describe possible association between the two variables. *For example, collect data from students in your class on whether or not they have a curfew on school nights and whether or not they have assigned chores at home. Is there evidence that those who have a curfew also tend to have chores?*

Foundational Standards

Solve real-life and mathematical problems using numerical and algebraic expressions and equations.

- Use variables to represent quantities in a real-world or mathematical problem, and construct simple equations and inequalities to solve problems by reasoning about the quantities.

Define, evaluate, and compare functions.

- Understand that a function is a rule that assigns to each input exactly one output. The graph of a function is the set of ordered pairs consisting of an input and the corresponding output.[2]

- Compare properties of two functions each represented in a different way (algebraically, graphically, numerically in tables, or by verbal descriptions). *For example, given a linear function represented by a table of values and a linear function represented by an algebraic expression, determine which function has the greater rate of change.*

---

[2] Function notation is not required in Grade 8.

Module 6: Linear Functions

- Interpret the equation $y = mx + b$ as defining a linear function, whose graph is a straight line; give examples of functions that are not linear. *For example, the function $A = s^2$ giving the area of a square as a function of its side length is not linear because its graph contains the points* (1,1)*,* (2,4) *and* (3,9)*, which are not on a straight line.*

Focus Standards for Mathematical Practice

- **Reason abstractly and quantitatively.** Students reason quantitatively by symbolically representing the verbal description of a relationship between two bivariate variables. They attend to the meaning of data based on the context of problems and the possible linear or nonlinear functions that explain the relationships of the variables.

- **Model with mathematics.** Students model relationships between variables using linear and nonlinear functions. They interpret models in the context of the data and reflect on whether or not the models make sense based on slopes, initial values, or the fit to the data.

- **Attend to precision.** Students evaluate functions to model a relationship between numerical variables. They evaluate the function by assessing the closeness of the data points to the line. They use care in interpreting the slope and the y-intercept in linear functions.

- **Look for and make use of structure.** Students identify pattern or structure in scatter plots. They fit lines to data displayed in a scatter plot and determine the equations of lines based on points or the slope and initial value.

Terminology

New or Recently Introduced Terms

- **Association (description)** (An *association* is a relationship between the two variables of a bivariate data set.

  The relationship is often expressed in terms of relative frequencies (described using two-way tables of the two domains of variables of the data set) or numerical relationships that can be modeled by functions (most often as linear relationships between the two domains of the two variables for the data set).)

- **Bivariate Data Set (description)** (A *bivariate data set* is an ordered list of ordered pairs of data values (called *data points*).

  Data sets and bivariate data sets are both called *data sets*. Data values can be either numerical or categorical. If both are numerical, then the data set is called a *numerical bivariate data set*.)

- **Column Relative Frequency** (In a two-way table, a *column relative frequency* is a cell frequency divided by the column total for that cell.)

- **Row Relative Frequency** (In a two-way table, a *row relative frequency* is the number given by dividing the cell frequency by the row total for that cell.)

- **Scatter Plot** (A *scatter plot* is a graph of a numerical bivariate data set.)

- **Two-Way Frequency Table (description)** (A *two-way frequency table* is a rectangular table used to summarize data on two categorical variables of a bivariate data set. The rows of the table correspond to the possible categories for one of the variables, and the columns correspond to the possible categories for the other variable. Entries in the cells of the table indicate the number of times that a particular category combination occurs in the data set; the value is the frequency for that combination.)
- **Variable (description)** (A *variable* is a symbol (such as a letter) that is a placeholder for a data value from a specified set of data values. The specified set of data values is called the *domain* of the variable.)

Familiar Terms and Symbols[3]

- Categorical variable
- Intercept or initial value
- Numerical variable
- Slope

Suggested Tools and Representations

- Graphing calculator
- Scatter Plot
- Two-way frequency tables

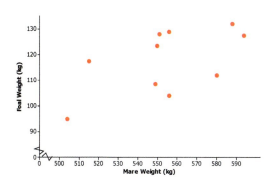

	Curfew	No Curfew	Total
Assigned Chores	25	10	35
Not Assigned Chores	8	7	15
Total	33	17	50

Scatter Plot   **Two-Way Frequency Table**

---

[3] These are terms and symbols students have seen previously.

Assessment Summary

Assessment Type	Administered	Format
Mid-Module Assessment Task	After Topic B	Constructed response with rubric
End-of-Module Assessment Task	After Topic D	Constructed response with rubric

A STORY OF RATIOS

Mathematics Curriculum

GRADE 8 • MODULE 6

Topic A
Linear Functions

Focus Standards:	■ Construct a function to model a linear relationship between two quantities. Determine the rate of change and initial value of the function from a description of a relationship or from two (x,y) values, including reading these from a table or from a graph. Interpret the rate of change and initial value of a linear function in terms of the situation it models, and in terms of its graph or a table of values.
	■ Describe qualitatively the functional relationship between two quantities by analyzing a graph (e.g., where the function is increasing or decreasing, linear or nonlinear). Sketch a graph that exhibits the qualitative features of a function that has been described verbally.
Instructional Days:	5
Lesson 1:	Modeling Linear Relationships (P)[1]
Lesson 2:	Interpreting Rate of Change and Initial Value (P)
Lesson 3:	Representations of a Line (P)
Lessons 4–5:	Increasing and Decreasing Functions (P, P)

In Topic A, students build on their study of functions by recognizing a linear relationship between two variables. Students use the context of a problem to construct a function to model a linear relationship. In Lesson 1, students are given a verbal description of a linear relationship between two variables and then must describe a linear model. Students graph linear functions using a table of values and by plotting points. They recognize a linear function given in terms of the slope and initial value, or y-intercept. In Lesson 2, students interpret the rate of change and the y-intercept, or initial value, in the context of the problem. They interpret the sign of the rate of change as indicating that a linear function is increasing or decreasing and as indicating the steepness of a line. In Lesson 3, students graph the line of a given linear function. They express the equation of a linear function as $y = mx + b$, or an equivalent form, when given the initial value and slope. In Lessons 4 and 5, students describe and interpret a linear function given two points or its graph.

---
[1]Lesson Structure Key: **P**-Problem Set Lesson, **M**-Modeling Cycle Lesson, **E**-Exploration Lesson, **S**-Socratic Lesson

Topic A: Linear Functions

A STORY OF RATIOS   Lesson 1   8•6

Lesson 1: Modeling Linear Relationships

Student Outcomes

- Students determine a linear function given a verbal description of a linear relationship between two quantities.
- Students interpret linear functions based on the context of a problem.
- Students sketch the graph of a linear function by constructing a table of values, plotting points, and connecting points by a line.

Lesson Notes

In this first lesson, students construct linear functions based on verbal descriptions of bivariate data. They graph the linear functions by creating a table of values, plotting points, and drawing the line. Throughout this lesson, provide students with the opportunity to explain the functions in terms of the equation of the line and the relationship between the two variables. Emphasize the context with students as they explain the rates of change and the initial values.

Classwork

Example 1 (2–3 minutes): Logging On

Read through the example as a class. Convey to students that the information presented in the example can be organized into ordered pairs or points. Minutes can be represented by x and cost by y.

> **Example 1: Logging On**
>
> Lenore has just purchased a tablet computer, and she is considering purchasing an Internet access plan so that she can connect to the Internet wirelessly from virtually anywhere in the world. One company offers an Internet access plan so that when a person connects to the company's wireless network, the person is charged a fixed access fee for connecting *plus* an amount for the number of minutes connected based upon a constant usage rate in dollars per minute.
>
> Lenore is considering this company's plan, but the company's advertisement does not state how much the fixed access fee for connecting is, nor does it state the usage rate. However, the company's website says that a 10-minute session costs $0.40, a 20-minute session costs $0.70, and a 30-minute session costs $1.00. Lenore decides to use these pieces of information to determine both the fixed access fee for connecting and the usage rate.

Exercises 1–6 (10 minutes)

This exercise set introduces students to constant rate of change and initial value and how those values are used to construct a function to model a situation. Pose each exercise to the class, one at a time, using the following questions to encourage discussion.

After Exercise 1, discuss as a class the need to graph this real-world problem only in the first quadrant. Begin by asking students the following:

- If we used the entire coordinate plane to graph this line, what would the negative x values represent?
  - *The x-axis represents minutes. So, time would be negative, which does not make sense in the context of the problem.*

A STORY OF RATIOS   Lesson 1   8•6

For Exercise 2, use the table to demonstrate constant rate of change to students.

- How could we use the table to determine the constant rate of change?

Number of Minutes	Total Session Cost in Dollars
0	
10	0.40
20	0.70
30	1.00
40	
50	
60	

(Between 10 and 20: +0.30; between 20 and 30: +0.30)

Exercises 1–6

1. Lenore makes a table of this information and a graph where number of minutes is represented by the horizontal axis and total session cost is represented by the vertical axis. Plot the three given points on the graph. These three points appear to lie on a line. What information about the access plan suggests that the correct model is indeed a linear relationship?

   The amount charged for the minutes connected is based upon a constant usage rate in dollars per minute.

Number of Minutes	Total Session Cost in Dollars
0	
10	0.40
20	0.70
30	1.00
40	
50	
60	

   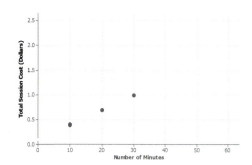

2. The rate of change describes how the total cost changes with respect to time.

   a. When the number of minutes increases by 10 (e.g., from 10 minutes to 20 minutes or from 20 minutes to 30 minutes), how much does the charge increase?

   When the number of minutes increases by 10 (e.g., from 10 minutes to 20 minutes or from 20 minutes to 30 minutes), the cost increases by $0.30 (30 cents).

   b. Another way to say this would be the usage charge per 10 minutes of use. Use that information to determine the increase in cost based on only 1 minute of additional usage. In other words, find the usage charge per minute of use.

   If $0.30 is the usage charge per 10 minutes of use, then $0.03 is the usage charge per 1 minute of use (i.e., the usage rate). Since the usage rate is constant, students should use what they have learned in Module 4.

   Lesson 1:   Modeling Linear Relationships

3. The company's pricing plan states that the usage rate is constant for any number of minutes connected to the Internet. In other words, the increase in cost for 10 more minutes of use (the value that you calculated in Exercise 2) is the same whether you increase from 20 to 30 minutes, 30 to 40 minutes, etc. Using this information, determine the total cost for 40 minutes, 50 minutes, and 60 minutes of use. Record those values in the table, and plot the corresponding points on the graph in Exercise 1.

Consider the following table and graphs.

Number of Minutes	Total Session Cost (in dollars)
0	
10	0.40
20	0.70
30	1.00
40	1.30
50	1.60
60	1.90

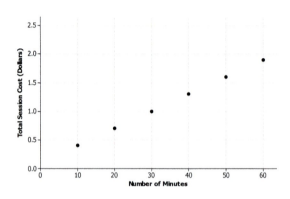

4. Using the table and the graph in Exercise 1, compute the hypothetical cost for 0 minutes of use. What does that value represent in the context of the values that Lenore is trying to figure out?

Since there is a $0.30 decrease in cost for each decrease of 10 minutes of use, one could subtract $0.30 from the cost value for 10 minutes and arrive at the hypothetical cost value for 0 minutes. That cost would be $0.10. Students may notice that such a value follows the regular pattern in the table and would represent the fixed access fee for connecting. (This value could also be found from the graph after completing Exercise 6.)

Convey to students that this is known as the initial value.

- Why is this a hypothetical cost?
  - *Because it is impossible to connect for 0 minutes; the connection is always for some interval of time.*

5. On the graph in Exercise 1, draw a line through the points representing 0 to 60 minutes of use under this company's plan. The slope of this line is equal to the constant rate of change, which in this case is the usage rate.

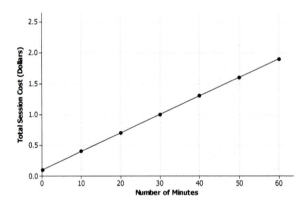

6. Using x for the number of minutes and y for the total cost in dollars, write a function to model the linear relationship between minutes of use and total cost.

$y = 0.03x + 0.10$

A STORY OF RATIOS   Lesson 1 8•6

Example 2 (2–3 minutes): Another Rate Plan

Provide students time to read the example. As a whole group, summarize this alternative rate plan.

> **Example 2: Another Rate Plan**
>
> A second wireless access company has a similar method for computing its costs. Unlike the first company that Lenore was considering, this second company explicitly states its access fee is 0.15, and its usage rate is 0.04 per minute.
>
> Total Session Cost = $\$0.15 + \$0.04$ (number of minutes)

- How is this plan presented differently?
  - *In this case, we are given the access fee and usage rate with an equation. In the first example, just data points were given.*
- Based on the work with the first set of problems, how do you think the two plans are different?
  - *The values for the access fee and usage charge per minute are different, or the initial value and the rate of change are different.*

Exercises 7–9 (7 minutes)

Allow students to work independently on these exercises. After most students have completed the problems, discuss them as a whole group.

> **Exercises 7–16**
>
> 7. Let x represent the number of minutes used and y represent the total session cost in dollars. Construct a linear function that models the total session cost based on the number of minutes used.
>
> $y = 0.04x + 0.15$
>
> 8. Using the linear function constructed in Exercise 7, determine the total session cost for sessions of 0, 10, 20, 30, 40, 50, and 60 minutes, and fill in these values in the table below.
>
> | Number of Minutes | Total Session Cost (in dollars) |
> |---|---|
> | 0 | 0.15 |
> | 10 | 0.55 |
> | 20 | 0.95 |
> | 30 | 1.35 |
> | 40 | 1.75 |
> | 50 | 2.15 |
> | 60 | 2.55 |

Lesson 1: Modeling Linear Relationships   11

9. Plot these points on the original graph in Exercise 1, and draw a line through these points. In what ways does the line that represents this second company's access plan differ from the line that represents the first company's access plan?

The second company's plan line begins at a greater initial value. The same plan also increases in total cost more quickly over time; in other words, the slope of the line for the second company's plan is steeper.

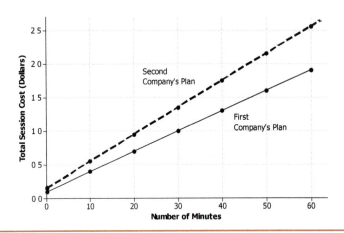

Exercises 10–12 (7 minutes)

MP3 download sites are a popular forum for selling music. Different sites offer pricing that depends on whether or not you want to purchase an entire album or individual songs à la carte. One site offers MP3 downloads of individual songs with the following price structure: a $3 fixed fee for a monthly subscription *plus* a charge of $0.25 per song.

10. Using x for the number of songs downloaded and y for the total monthly cost in dollars, construct a linear function to model the relationship between the number of songs downloaded and the total monthly cost.

    Since $3 is the initial cost and there is a 25 cent increase per song, the function would be

    $y = 3 + 0.25x$ or $y = 0.25x + 3$.

11. Using the linear function you wrote in Exercise 10, construct a table to record the total monthly cost (in dollars) for MP3 downloads of 10 songs, 20 songs, and so on up to 100 songs.

Number of Songs	Total Monthly Cost (in dollars)
10	5.50
20	8.00
30	10.50
40	13.00
50	15.50
60	18.00
70	20.50
80	23.00
90	25.50
100	28.00

12. Plot the 10 data points in the table on a coordinate plane. Let the x-axis represent the number of songs downloaded and the y-axis represent the total monthly cost (in dollars) for MP3 downloads.

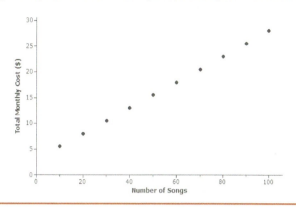

Exercises 13–16 (7–8 minutes)

Read through the problem as a class. The data in this exercise set are presented as two points given in context. Point out the difference by asking students the following:

- How are the data in this problem different from the data in the MP3 problem?
  - In this problem, the data can be organized as ordered pairs. In the MP3 problem, a rate of change and initial value were given.

A band will be paid a flat fee for playing a concert. Additionally, the band will receive a fixed amount for every ticket sold. If 40 tickets are sold, the band will be paid $200. If 70 tickets are sold, the band will be paid $260.

13. Determine the rate of change.

    The points $(40, 200)$ and $(70, 260)$ have been given.

    So, the rate of change is 2 because $\dfrac{260 - 200}{70 - 40} = 2$.

14. Let x represent the number of tickets sold and y represent the amount the band will be paid in dollars. Construct a linear function to represent the relationship between the number of tickets sold and the amount the band will be paid.

    Using the rate of change and $(40, 200)$:

    $200 = 2(40) + b$
    $200 = 80 + b$
    $120 = b$

    Therefore, the function is $y = 2x + 120$.

15. What flat fee will the band be paid for playing the concert regardless of the number of tickets sold?

    The band will be paid a flat fee of $120 for playing the concert.

16. How much will the band receive for each ticket sold?

    The band receives $2 per ticket.

Lesson 1: Modeling Linear Relationships

Closing (5 minutes)

Consider posing the following questions; allow a few student responses for each.

- In Exercise 9 when the two pricing models that Lenore was considering were both displayed on the same graph, was there ever a point at which the second company's model was a better, less expensive choice than the first company's model?
    - *No. The second company always had the more expensive plan; its line was always above the other company's line.*
- When comparing the equations of the two models, which value in the second company's model (the $0.15 access fee or $0.04 cost per minute) led you to think that it would increase at a faster rate than the first model?
    - *The $0.04 cost per minute led me to believe it would increase at a faster rate. The other company's plan only increased at a rate of $0.03 per minute.*

Lesson Summary

A linear function can be used to model a linear relationship between two types of quantities. The graph of a linear function is a straight line.

A linear function can be constructed using a rate of change and an initial value. It can be interpreted as an equation of a line in which:

- The rate of change is the slope of the line and describes how one quantity changes with respect to another quantity.
- The initial value is the y-intercept.

Exit Ticket (5 minutes)

Name _____ Date _____

Lesson 1: Modeling Linear Relationships

Exit Ticket

A rental car company offers a rental package for a midsize car. The cost comprises a fixed $30 administrative fee for the cleaning and maintenance of the car plus a rental cost of $35 per day.

1. Using x for the number of days and y for the total cost in dollars, construct a function to model the relationship between the number of days and the total cost of renting a midsize car.

2. The same company is advertising a deal on compact car rentals. The linear function $y = 30x + 15$ can be used to model the relationship between the number of days, x, and the total cost in dollars, y, of renting a compact car.

    a. What is the fixed administrative fee?

    b. What is the rental cost per day?

Exit Ticket Sample Solutions

A rental car company offers a rental package for a midsize car. The cost comprises a fixed $30 administrative fee for the cleaning and maintenance of the car plus a rental cost of $35 per day.

1. Using x for the number of days and y for the total cost in dollars, construct a function to model the relationship between the number of days and the total cost of renting a midsize car.

   $y = 35x + 30$

2. The same company is advertising a deal on compact car rentals. The linear function $y = 30x + 15$ can be used to model the relationship between the number of days, x, and the total cost in dollars, y, of renting a compact car.

   a. What is the fixed administrative fee?

   The administrative fee is $15.

   b. What is the rental cost per day?

   It costs $30 per day to rent the compact car.

Problem Set Sample Solutions

1. Recall that Lenore was investigating two wireless access plans. Her friend in Europe says that he uses a plan in which he pays a monthly fee of 30 euro plus 0.02 euro per minute of use.

   a. Construct a table of values for his plan's monthly cost based on 100 minutes of use for the month, 200 minutes of use, and so on up to 1,000 minutes of use. (The charge of 0.02 euro per minute of use is equivalent to 2 euro per 100 minutes of use.)

Number of Minutes	Total Monthly Cost (€)
100	32.00
200	34.00
300	36.00
400	38.00
500	40.00
600	42.00
700	44.00
800	46.00
900	48.00
1,000	50.00

b. Plot these 10 points on a carefully labeled graph, and draw the line that contains these points.

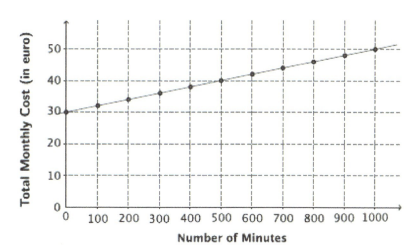

c. Let x represent minutes of use and y represent the total monthly cost in euro. Construct a linear function that determines the monthly cost based on minutes of use.

$y = 30 + 0.02x$

d. Use the function to calculate the cost under this plan for 750 minutes of use. If this point were added to the graph, would it be above the line, below the line, or on the line?

The cost for 750 minutes would be €45. The point (750, 45) would be on the line.

2. A shipping company charges a $4.45 handling fee in addition to $0.27 per pound to ship a package.

   a. Using x for the weight in pounds and y for the cost of shipping in dollars, write a linear function that determines the cost of shipping based on weight.

   $y = 4.45 + 0.27x$

b. Which line (solid, dotted, or dashed) on the following graph represents the shipping company's pricing method? Explain.

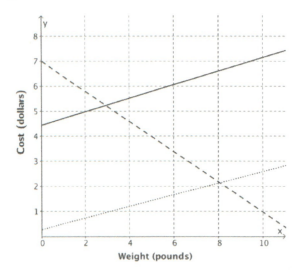

The solid line would be the correct line. Its initial value is 4.45, and its slope is 0.27. The dashed line shows the cost decreasing as the weight increases, so that is not correct. The dotted line starts at an initial value that is too low.

3. Kelly wants to add new music to her MP3 player. Another subscription site offers its downloading service using the following: Total Monthly Cost $= 5.25 + 0.30$(number of songs).

   a. Write a sentence (all words, no math symbols) that the company could use on its website to explain how it determines the price for MP3 downloads for the month.

   "We charge a $5.25 subscription fee plus 30 cents per song downloaded."

   b. Let x represent the number of songs downloaded and y represent the total monthly cost in dollars. Construct a function to model the relationship between the number of songs downloaded and the total monthly cost.

   $y = 5.25 + 0.30x$

   c. Determine the cost of downloading 10 songs.

   $5.25 + 0.30(10) = 8.25$

   The cost of downloading 10 songs is $8.25.

4. Li Na is saving money. Her parents gave her an amount to start, and since then she has been putting aside a fixed amount each week. After six weeks, Li Na has a total of $82 of her own savings in addition to the amount her parents gave her. Fourteen weeks from the start of the process, Li Na has $118.

   a. Using x for the number of weeks and y for the amount in savings (in dollars), construct a linear function that describes the relationship between the number of weeks and the amount in savings.

   The points $(6, 82)$ and $(14, 118)$ have been given.

   So, the rate of change is 4.5 because $\frac{118 - 82}{14 - 6} = \frac{36}{8} = 4.5$.

   Using the rate of change and $(6, 82)$:

   $82 = 4.5(6) + b$

   $82 = 27 + b$

   $55 = b$

   The function is $y = 4.5x + 55$.

   b. How much did Li Na's parents give her to start?

   Li Na's parents gave her $55 to start.

   c. How much does Li Na set aside each week?

   Li Na is setting aside $4.50 every week for savings.

   d. Draw the graph of the linear function below (start by plotting the points for $x = 0$ and $x = 20$).

   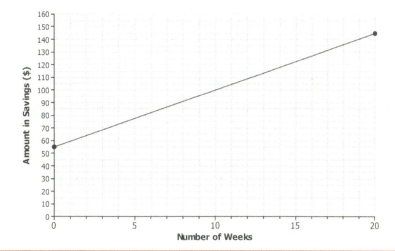

Lesson 1: Modeling Linear Relationships

Lesson 2: Interpreting Rate of Change and Initial Value

Student Outcomes

- Students interpret the constant rate of change and initial value of a line in context.
- Students interpret slope as rate of change and relate slope to the steepness of a line and the sign of the slope, indicating that a linear function is increasing if the slope is positive and decreasing if the slope is negative.

Lesson Notes

In this lesson, students work with linear functions and the equations that define the linear functions. They specifically interpret the slope of a linear function as a rate of change. *Rate of change* is used to mean *constant rate of change* in the subsequent lessons. Students also explain whether the rate of change of a linear function is increasing or decreasing. Each example in this lesson has a context. Connect students to the context of each problem by having them summarize what they think a function indicates about the problem. For example, have students explain a slope in terms of the units and the rate of change. Also, ask students to explain how knowing the value of one of the variables predicts the value of the second variable.

Classwork

> Linear functions are defined by the equation of a line. The graphs and the equations of the lines are important for understanding the relationship between the two variables represented in the following example as x and y.

Example 1 (5 minutes): Rate of Change and Initial Value

Read through the site's pricing plan. Convey to students that the rate of change and initial value can immediately be found when given an equation written in the form $y = mx + b$ or $y = a + bx$. Pay careful attention to the interpretation of the rate of change and initial value. Give students a moment to answer parts (a) and (b) individually. Then, discuss as a class the interpretation of rate of change and initial value in context, and generalize the interpretation of rate of change and initial value in contextual situations. Discuss why the sign of the rate of change affects whether or not the linear function increases or decreases.

> **Example 1: Rate of Change and Initial Value**
>
> The equation of a line can be interpreted as defining a linear function. The graphs and the equations of lines are important in understanding the relationship between two types of quantities (represented in the following examples by x and y).
>
> In a previous lesson, you encountered an MP3 download site that offers downloads of individual songs with the following price structure: a $3 fixed fee for a monthly subscription *plus* a fee of $0.25 per song. The linear function that models the relationship between the number of songs downloaded and the total monthly cost of downloading songs can be written as
>
> $$y = 0.25x + 3,$$
>
> where x represents the number of songs downloaded and y represents the total monthly cost (in dollars) for MP3 downloads.

a. In your own words, explain the meaning of 0.25 within the context of the problem.

In the example on the previous page, the value 0.25 means there is a cost increase of $\$0.25$ for every 1 song downloaded.

b. In your own words, explain the meaning of 3 within the context of the problem.

In the example on the previous page, the value of 3 represents an initial cost of $\$3$ for downloading 0 songs. In other words, there is a fixed cost of $\$3$ to subscribe to the site.

The values represented in the function can be interpreted in the following way:

$$y = \underbrace{0.25x}_{\text{rate of change}} + \underbrace{3}_{\text{initial value}}$$

The coefficient of x is referred to as the *rate of change*. It can be interpreted as the change in the values of y for every one-unit increase in the values of x. When the rate of change is positive, the linear function is *increasing*. In other words, *increasing* indicates that as the x-value increases, so does the y-value. When the rate of change is negative, the linear function is *decreasing*. *Decreasing* indicates that as the x-value increases, the y-value decreases.	The constant value is referred to as the *initial value* or y-intercept and can be interpreted as the value of y when $x = 0$.

Exercises 1–6 (15 minutes): Is It a Better Deal?

Discuss the other site's pricing plan. This second plan results in a different linear function to determine pricing. Let students work independently, and then discuss as a class the linear function and compare it to the first company that is summarized in the lesson.

Exercises 1–6: Is It a Better Deal?

Another site offers MP3 downloads with a different price structure: a $\$2$ fixed fee for a monthly subscription *plus* a fee of $\$0.40$ per song.

1. Write a linear function to model the relationship between the number of songs downloaded and the total monthly cost. As before, let x represent the number of songs downloaded and y represent the total monthly cost (in dollars) of downloading songs.

   $y = 0.4x + 2$

2. Determine the cost of downloading 0 songs and 10 songs from this site.

   $y = 0.4(0) + 2 = 2.00$. *For 0 songs, the cost is $\$2.00$.*

   $y = 0.4(10) + 2 = 6.00$. *For 10 songs, the cost is $\$6.00$.*

Lesson 2: Interpreting Rate of Change and Initial Value

3. The graph below already shows the linear model for the first subscription site (Company 1): $y = 0.25x + 3$. Graph the equation of the line for the second subscription site (Company 2) by marking the two points from your work in Exercise 2 (for 0 songs and 10 songs) and drawing a line through those two points.

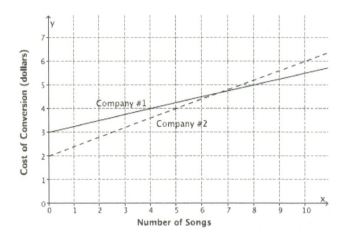

4. Which line has a steeper slope? Which company's model has the more expensive cost per song?

   The line modeled by the second subscription site (Company 2) is steeper. It has the larger slope value and the greater cost per song.

5. Which function has the greater initial value?

   The first subscription site (Company 1) has the greater initial value. Its monthly subscription fee is $3 compared to only $2 for the second site.

6. Which subscription site would you choose if you only wanted to download 5 songs per month? Which company would you choose if you wanted to download 10 songs? Explain your reasoning.

   For 5 songs: Company 1's cost is $4.25 ($y = 25(5) + 3$); Company 2's cost is $4.00 ($y = 0.4(5) + 2$). So, Company 2 would be the better choice. Graphically, Company 2's model also has the smaller y-value when $x = 5$.

   For 10 songs: Company 1's cost is $5.50 ($y = 0.25(10) + 3$); Company 2's cost is $6.00 ($y = 0.4(10) + 2$). So, Company 1 would be the better choice. Graphically, Company 1's model also has the smaller y-value at $x = 10$.

A STORY OF RATIOS   Lesson 2   8•6

Exercises 7–9 (10–15 minutes): Aging Autos

Let students work independently, and then discuss the answers as a class. Note the linear equation provided in Exercise 8 is written in the form $y = a + bx$. Students may mix up the values for rate of change and initial value. If class time is running short, choose two of the exercises for students to work on, and assign the other exercise of the Problem Set for homework.

Exercises 7–9: Aging Autos

7. When someone purchases a new car and begins to drive it, the mileage (meaning the number of miles the car has traveled) immediately increases. Let x represent the number of years since the car was purchased and y represent the total miles traveled. The linear function that models the relationship between the number of years since purchase and the total miles traveled is $y = 15000x$.

   a. Identify and interpret the rate of change.

   The rate of change is $15,000$. It means that the mileage is increasing by $15,000$ miles per year.

   b. Identify and interpret the initial value.

   The initial value is 0. This means that there were no miles on the car when it was purchased.

   c. Is the mileage increasing or decreasing each year according to the model? Explain your reasoning.

   Since the rate of change is positive, it means the mileage is increasing each year.

8. When someone purchases a new car and begins to drive it, generally speaking, the resale value of the car (in dollars) goes down each year. Let x represent the number of years since purchase and y represent the resale value of the car (in dollars). The linear function that models the resale value based on the number of years since purchase is $y = 20000 - 1200x$.

   a. Identify and interpret the rate of change.

   The rate of change is $-1,200$. The resale value of the car is decreasing $\$1,200$ every year since purchase.

   b. Identify and interpret the initial value.

   The initial value is $\$20,000$. The car's value at the time of purchase was $\$20,000$.

   c. Is the resale value increasing or decreasing each year according to the model? Explain.

   The slope is negative. This means that the resale value decreases each year.

9. Suppose you are given the linear function $y = 2.5x + 10$.

   a. Write a story that can be modeled by the given linear function.

   Answers will vary. I am ordering cupcakes for a birthday party. The bakery is going to charge $\$2.50$ per cupcake in addition to a $\$10$ decorating fee.

   b. What is the rate of change? Explain its meaning with respect to your story.

   The rate of change is 2.5, which means that the cost increases $\$2.50$ for every additional cupcake ordered.

   c. What is the initial value? Explain its meaning with respect to your story.

   The initial value is 10, which in this story means that there is a flat fee of $\$10$ to decorate the cupcakes.

Lesson 2:   Interpreting Rate of Change and Initial Value   23

©2018 Great Minds®. eureka-math.org

Closing (5 minutes)

Consider posing the following questions. Allow a few student responses for each.

- In Exercise 3, for what number of songs would the total monthly cost be the same regardless of the company selected? What visual attribute of the graph supports this answer?
  - *7 songs; point of intersection*

It may be necessary to discuss why the answer is 7 and not $6\frac{2}{3}$ (the solution you would get if you solved the system algebraically).

- Just by looking at the graph for Exercise 3, which company would you select if you had 12 songs to download? Explain why this is the better choice.
  - *Company 1 has the lower cost for more than 7 songs since its linear model is below the Company 2 linear model after 7 songs.*

Lesson Summary

When a linear function is given by the equation of a line of the form $y = mx + b$, the rate of change is m, and the initial value is b. Both are easy to identify.

The rate of change of a linear function is the slope of the line it represents. It is the change in the values of y per a one-unit increase in the values of x.

- A positive rate of change indicates that a linear function is increasing.
- A negative rate of change indicates that a linear function is decreasing.
- Given two lines each with positive slope, the function represented by the steeper line has a greater rate of change.

The initial value of a linear function is the value of the y-variable when the x-value is zero.

Exit Ticket (10 minutes)

A STORY OF RATIOS   Lesson 2   8•6

Name _____   Date _____

Lesson 2: Interpreting Rate of Change and Initial Value

Exit Ticket

In 2008, a collector of sports memorabilia purchased 5 specific baseball cards as an investment. Let y represent each card's resale value (in dollars) and x represent the number of years since purchase. Each card's resale value after 0, 1, 2, 3, and 4 years could be modeled by linear equations as follows:

Card A: $y = 5 - 0.7x$

Card B: $y = 4 + 2.6x$

Card C: $y = 10 + 0.9x$

Card D: $y = 10 - 1.1x$

Card E: $y = 8 + 0.25x$

1. Which card(s) are decreasing in value each year? How can you tell?

2. Which card(s) had the greatest initial value at purchase (at 0 years)?

3. Which card(s) is increasing in value the fastest from year to year? How can you tell?

4. If you were to graph the equations of the resale values of Card B and Card C, which card's graph line would be steeper? Explain.

5. Write a sentence explaining the 0.9 value in Card C's equation.

Exit Ticket Sample Solutions

In 2008, a collector of sports memorabilia purchased 5 specific baseball cards as an investment. Let y represent each card's resale value (in dollars) and x represent the number of years since purchase. Each card's resale value after 0, 1, 2, 3, and 4 years could be modeled by linear equations as follows:

Card A: $y = 5 - 0.7x$

Card B: $y = 4 + 2.6x$

Card C: $y = 10 + 0.9x$

Card D: $y = 10 - 1.1x$

Card E: $y = 8 + 0.25x$

1. Which card(s) are decreasing in value each year? How can you tell?

   Cards A and D are decreasing in value, as shown by the negative values for rate of change in each equation.

2. Which card(s) had the greatest initial value at purchase (at 0 years)?

   Since all of the models are in slope-intercept form, Cards C and D have the greatest initial values at $10 each.

3. Which card(s) is increasing in value the fastest from year to year? How can you tell?

   Card B is increasing in value the fastest from year to year. Its model has the greatest rate of change.

4. If you were to graph the equations of the resale values of Card B and Card C, which card's graph line would be steeper? Explain.

   The Card B line would be steeper because the function for Card B has the greatest rate of change; the card's value is increasing at a faster rate than the other values of other cards.

5. Write a sentence explaining the 0.9 value in Card C's equation.

   The 0.9 value means that Card C's value increases by 90 cents per year.

Problem Set Sample Solutions

1. A rental car company offers the following two pricing methods for its customers to choose from for a one-month rental:

   Method 1: Pay $400 for the month, or

   Method 2: Pay $0.30 per mile plus a standard maintenance fee of $35.

   a. Construct a linear function that models the relationship between the miles driven and the total rental cost for Method 2. Let x represent the number of miles driven and y represent the rental cost (in dollars).

   $y = 35 + 0.30x$

   b. If you plan to drive 1,100 miles for the month, which method would you choose? Explain your reasoning.

   Method 1 has a flat rate of $400 regardless of miles. Using Method 2, the cost would be $365 ($y = 35 + 0.3(1100)$). So, Method 2 would be preferred.

Lesson 2: Interpreting Rate of Change and Initial Value

2. Recall from a previous lesson that Kelly wants to add new music to her MP3 player. She was interested in a monthly subscription site that offered its MP3 downloading service for a monthly subscription fee *plus* a fee per song. The linear function that modeled the total monthly cost in dollars (y) based on the number of songs downloaded (x) is $y = 5.25 + 0.30x$.

The site has suddenly changed its monthly price structure. The linear function that models the new total monthly cost in dollars (y) based on the number of songs downloaded (x) is $y = 0.35x + 4.50$.

   a. Explain the meaning of the value 4.50 in the new equation. Is this a better situation for Kelly than before?

   The initial value is 4.50 and means that the monthly subscription cost is now $\$4.50$. This is lower than before, which is good for Kelly.

   b. Explain the meaning of the value 0.35 in the new equation. Is this a better situation for Kelly than before?

   The rate of change is 0.35. This means that the cost is increasing by $\$0.35$ for every song downloaded. This is more than the download cost for the original plan.

   c. If you were to graph the two equations (old versus new), which line would have the steeper slope? What does this mean in the context of the problem?

   The slope of the new line is steeper because the new linear function has a greater rate of change. It means that the total monthly cost of the new plan is increasing at a faster rate per song compared to the cost of the old plan.

   d. Which subscription plan provides the better value if Kelly downloads fewer than 15 songs per month?

   If Kelly were to download 15 songs, both plans will cost the same ($\$9.75$). Therefore, the new plan is cheaper if Kelly downloads fewer than 15 songs.

Lesson 2: Interpreting Rate of Change and Initial Value

A STORY OF RATIOS   Lesson 3   8•6

Lesson 3: Representations of a Line

Student Outcomes

- Students graph a line specified by a linear function.
- Students graph a line specified by an initial value and a rate of change of a function and construct the linear function by interpreting the graph.
- Students graph a line specified by two points of a linear relationship and provide the linear function.

Lesson Notes

Linear functions are defined by the equations of a line. This lesson reviews students' work with the representation of a line and, in particular, the determination of the rate of change and the initial value of a linear function from two points on the graph or from the equation of the line defined by the function in the form $y = mx + b$ or an equivalent form. Students interpret the rate of change and the initial value based on the graph of the equation of the line in addition to the context of the variables.

Classwork

Example 1 (10 minutes): Rate of Change and Initial Value Given in the Context of the Problem

Here, verbal information giving an initial value and a rate of change is translated into a function and its graph. Work through the example as a class.

In part (b), explain why the value 0.5 given in the question is the rate of change.

- It is a good idea to show this on the graph, demonstrating that each increase of 1 unit for m (miles) results in an increase of 0.5 for C (cost in dollars). An increase of 1,000 for m results in an increase of 500 units for C.
- Point out that if the question stated that each mile driven *reduced* the cost by $0.50, then the line would have a negative slope.

It is important for students to understand that when the scales on the two axes are different, the rate of change cannot be used to plot points by simply counting the squares. Encourage students to use the rate of change by holding on to the idea of increasing the variable shown on the horizontal axis and showing the resulting increase in the variable shown on the vertical axis (as explained in the previous paragraph).

Given the rate of change and initial value, the linear function can be written in slope-intercept form ($y = mx + b$) or an equivalent form such as $y = a + bx$. Students should pay careful attention to variables presented in the problem; m and C are used in place of x and y.

28   Lesson 3:   Representations of a Line

Example 1: Rate of Change and Initial Value Given in the Context of the Problem

A truck rental company charges a 150 rental fee in addition to a charge of $\$0.50$ per mile driven. Graph the linear function relating the total cost of the rental in dollars, C, to the number of miles driven, m, on the axes below.

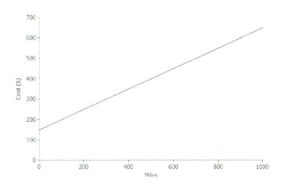

a. If the truck is driven 0 miles, what is the cost to the customer? How is this shown on the graph?

$\$150$, shown as the point $(0, 150)$. This is the initial value. Some students might say "b." Help them to use the term initial value.

b. What is the rate of change that relates cost to number of miles driven? Explain what it means within the context of the problem.

The rate of change is 0.5. It means that the cost increases by $\$0.50$ for every mile driven.

c. On the axes given, sketch the graph of the linear function that relates C to m.

Students can plot the initial value $(0, 150)$ and then use the rate of change to identify additional points as needed. A $1,000$-unit increase in m results in a 500-unit increase for C, so another point on the line is $(1000, 650)$.

d. Write the equation of the linear function that models the relationship between number of miles driven and total rental cost.

$C = 0.5m + 150$

Exercises 1–5 (10 minutes)

Here, students have an opportunity to practice the ideas to which they have just been introduced. Let students work independently on these exercises. Then, discuss and confirm answers as a class.

Exercise 3, part (c), provides an excellent opportunity for discussion about the model and whether or not it continues to make sense over time.

- In Exercise 3, you found that the price of the car in year seven was less than $600. Does this make sense in general?
    - *Not really*
- Under what conditions might the car be worth less than $600 after seven years?
    - *The car may have been in an accident or damaged.*

Lesson 3: Representations of a Line

Exercises

Jenna bought a used car for $18,000. She has been told that the value of the car is likely to decrease by $2,500 for each year that she owns the car. Let the value of the car in dollars be V and the number of years Jenna has owned the car be t.

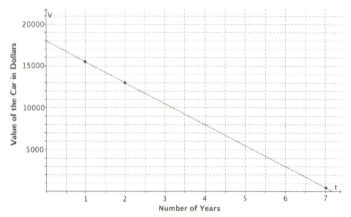

1. What is the value of the car when $t = 0$? Show this point on the graph.

   $18,000.$ *Shown by the point* $(0, 18000)$

2. What is the rate of change that relates V to t? (Hint: Is it positive or negative? How can you tell?)

   $-2,500.$ *The rate of change is negative because the value of the car is decreasing.*

3. Find the value of the car when:
   a. $t = 1$

   $\$18000 - \$2500 = \$15500$

   b. $t = 2$

   $\$18000 - 2(\$2500) = \$13000$

   c. $t = 7$

   $\$18000 - 7(\$2500) = \$500$

4. Plot the points for the values you found in Exercise 3, and draw the line (using a straightedge) that passes through those points.

   See the graph above.

5. Write the linear function that models the relationship between the number of years Jenna has owned the car and the value of the car.

   $V = 18000 - 2500t$ *or* $V = -2500t + 18000$

Exercises 6–10 (10 minutes)

Here, in the context of the pricing of a book, students are given two points on the graph of a linear equation and are expected to draw the graph, determine the rate of change, and answer questions by referring to the graph.

Point out that the horizontal axis does not start at 0. Ask students the following question:

- Why do you think the first value is at 15?
  - *The online bookseller may not sell the book for less than $15.*

In Exercise 8, students are asked to find the rate of change; it might be worthwhile to check that they are using the scales on the axes, not purely counting squares.

For Exercises 9 and 10, encourage students to show their work by drawing vertical and horizontal lines on the graph, as shown in the sample student answers below.

Let students work with a partner. Then, discuss and confirm answers as a class.

An online bookseller has a new book in print. The company estimates that if the book is priced at $15, then 800 copies of the book will be sold per day, and if the book is priced at $20, then 550 copies of the book will be sold per day.

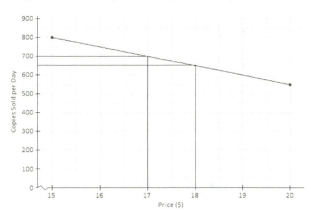

6. Identify the ordered pairs given in the problem. Then, plot both on the graph.

   The ordered pairs are $(15, 800)$ and $(20, 550)$. See the graph above.

7. Assume that the relationship between the number of books sold and the price is linear. (In other words, assume that the graph is a straight line.) Using a straightedge, draw the line that passes through the two points.

   See the graph above.

8. What is the rate of change relating number of copies sold to price?

   Between the points $(15, 800)$ and $(20, 550)$, the run is 5, and the rise is $-(800 - 550) = -250$. So, the rate of change is $\frac{-250}{5} = -50$.

9. Based on the graph, if the company prices the book at $18, about how many copies of the book can they expect to sell per day?

   650

10. Based on the graph, approximately what price should the company charge in order to sell 700 copies of the book per day?

    $17

Lesson 3: Representations of a Line

Closing (5 minutes)

If time allows, consider posing the following questions:

- How would you interpret the meaning of the rate of change (-50) from Exercise 8?
  - *Answers will vary; pay careful attention to wording. The number of copies sold would decrease by 50 units as the price increased by $1, or for every dollar increase in the price, the number of copies sold would decrease by 50 units.*
- Does it seem reasonable that the number of copies sold would decrease with respect to an increase in price?
  - *Yes, if the book is really expensive, someone may not want to buy it. If the cost remains low, it seems reasonable that more people would want to buy it.*
- How is the information given in the truck rental problem different from the information given in the book-pricing problem?
  - *In the book pricing problem, the information is given as ordered pairs. In the truck rental problem, the information is given in the form of a slope and an initial value.*

Lesson Summary

When the rate of change, b, and an initial value, a, are given in the context of a problem, the linear function that models the situation is given by the equation $y = a + bx$.

The rate of change and initial value can also be used to sketch the graph of the linear function that models the situation.

When two or more ordered pairs are given in the context of a problem that involves a linear relationship, the graph of the linear function is the line that passes through those points. The linear function can be represented by the equation of that line.

Exit Ticket (10 minutes)

Name _____ Date_____

Lesson 3: Representations of a Line

Exit Ticket

1. A car starts a journey with 18 gallons of fuel. Assuming a constant rate, the car consumes 0.04 gallon for every mile driven. Let A represent the amount of gas in the tank (in gallons) and m represent the number of miles driven.

   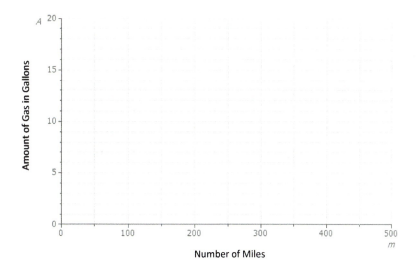

   a. How much gas is in the tank if 0 miles have been driven? How would this be represented on the axes above?

   b. What is the rate of change that relates the amount of gas in the tank to the number of miles driven? Explain what it means within the context of the problem.

   c. On the axes above, draw the line that represents the graph of the linear function that relates A to m.

   d. Write the linear function that models the relationship between the number of miles driven and the amount of gas in the tank.

2. Andrew works in a restaurant. The graph below shows the relationship between the amount Andrew earns in dollars and the number of hours he works.

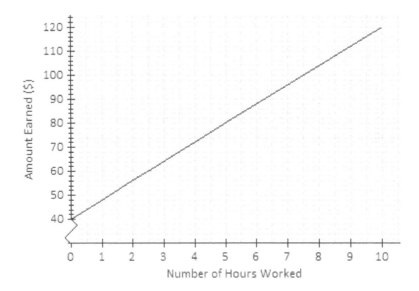

a. If Andrew works for 7 hours, approximately how much does he earn in dollars?

b. Estimate how long Andrew has to work in order to earn $64.

c. What is the rate of change of the function given by the graph? Interpret the value within the context of the problem.

Exit Ticket Sample Solutions

1. A car starts a journey with 18 gallons of fuel. Assuming a constant rate, the car consumes 0.04 gallon for every mile driven. Let A represent the amount of gas in the tank (in gallons) and m represent the number of miles driven.

    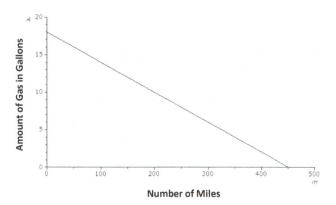

    a. How much gas is in the tank if 0 miles have been driven? How would this be represented on the axes above?

    There are 18 gallons in the tank. This would be represented as $(0, 18)$, the initial value, on the graph above.

    b. What is the rate of change that relates the amount of gas in the tank to the number of miles driven? Explain what it means within the context of the problem.

    -0.04; the car consumes 0.04 gallon for every mile driven. It relates the amount of fuel to the miles driven.

    c. On the axes above, draw the line that represents the graph of the linear function that relates A to m.

    See the graph above. Students can plot the initial value $(0, 18)$ and then use the rate of change to identify additional points as needed. A 50-unit increase in m results in a 2-unit decrease for A, so another point on the line is $(50, 16)$.

    d. Write the linear function that models the relationship between the number of miles driven and the amount of gas in the tank.

    $A = 18 - 0.04m$ or $A = -0.04m + 18$

2. Andrew works in a restaurant. The graph below shows the relationship between the amount Andrew earns in dollars and the number of hours he works.

    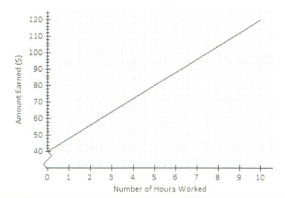

Lesson 3: Representations of a Line

a. If Andrew works for 7 hours, approximately how much does he earn in dollars?

$96

b. Estimate how long Andrew has to work in order to earn $64.

3 hours

c. What is the rate of change of the function given by the graph? Interpret the value within the context of the problem.

Using the ordered pairs $(7, 96)$ and $(3, 64)$, the slope is 8. It means that the amount Andrew earns increases by $8 for every hour worked.

Problem Set Sample Solutions

1. A plumbing company charges a service fee of $120, plus $40 for each hour worked. Sketch the graph of the linear function relating the cost to the customer (in dollars), C, to the time worked by the plumber (in hours), t, on the axes below.

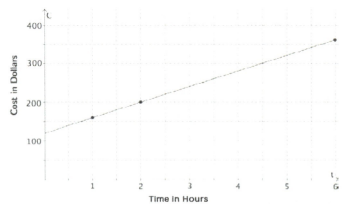

a. If the plumber works for 0 hours, what is the cost to the customer? How is this shown on the graph?

$120 This is shown on the graph by the point $(0, 120)$.

b. What is the rate of change that relates cost to time?

40

c. Write a linear function that models the relationship between the hours worked and the cost to the customer.

$C = 40t + 120$

d. Find the cost to the customer if the plumber works for each of the following number of hours.

 i. 1 hour

 $160

ii. 2 hours

$200

iii. 6 hours

$360

e. Plot the points for these times on the coordinate plane, and use a straightedge to draw the line through the points.

See the graph on the previous page.

2. An author has been paid a writer's fee of $1,000 plus $1.50 for every copy of the book that is sold.

a. Sketch the graph of the linear function that relates the total amount of money earned in dollars, A, to the number of books sold, n, on the axes below.

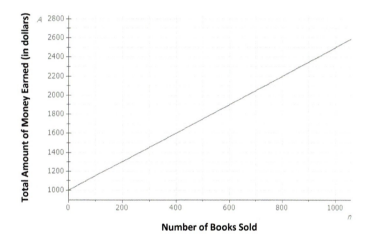

b. What is the rate of change that relates the total amount of money earned to the number of books sold?

1.5

c. What is the initial value of the linear function based on the graph?

$1,000$

d. Let the number of books sold be n and the total amount earned be A. Construct a linear function that models the relationship between the number of books sold and the total amount earned.

$A = 1.5n + 1000$

3. Suppose that the price of gasoline has been falling. At the beginning of last month ($t = 0$), the price was $4.60 per gallon. Twenty days later ($t = 20$), the price was $4.20 per gallon. Assume that the price per gallon, P, fell at a constant rate over the twenty days.

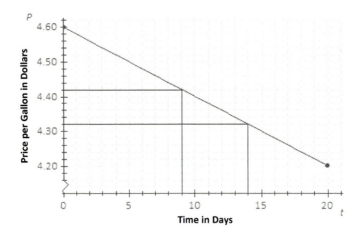

a. Identify the ordered pairs given in the problem. Plot both points on the coordinate plane above.

$(0, 4.60)$ and $(20, 4.20)$; *see the graph above.*

b. Using a straightedge, draw the line that contains the two points.

See the graph above.

c. What is the rate of change? What does it mean within the context of the problem?

Using points $(0, 4.60)$ and $(20, 4.20)$, the rate of change is -0.02 because $\frac{4.20 - 4.60}{20 - 0} = \frac{-0.4}{20} = -0.02$. The price of gas is decreasing $0.02 each day.

d. What is the function that models the relationship between the number of days and the price per gallon?

$P = -0.02t + 4.6$

e. What was the price of gasoline after 9 days?

$4.42; *see the graph above.*

f. After how many days was the price $4.32?

14 days; see the graph above.

Lesson 4: Increasing and Decreasing Functions

Student Outcomes

- Students describe qualitatively the functional relationship between two types of quantities by analyzing a graph.
- Students sketch a graph that exhibits the qualitative features of a function based on a verbal description.

Lesson Notes

This lesson focuses on graphs and the role they play in analyzing functional relationships between quantities. Students have been introduced to increasing and decreasing functions in a prior lesson in Grade 8. This lesson references a constant function, one in which the graph of the function is a line with zero slope. Piecewise functions are also used throughout the lesson to demonstrate how the functional relationship can increase or decrease between different intervals. Rate of change should be discussed among the intervals, but the term *piecewise function* does not need to be defined. This lesson also focuses on linear relationships. Nonlinear examples are presented in the next lesson.

Classwork

Opening

> Graphs are useful tools in terms of representing data. They provide a visual story, highlighting important facts that surround the relationship between quantities.
>
> The graph of a linear function is a line. The slope of the line can provide useful information about the functional relationship between the two types of quantities:
>
> - A linear function whose graph has a positive slope is said to be an *increasing function*.
> - A linear function whose graph has a negative slope is said to be a *decreasing function*.
> - A linear function whose graph has a zero slope is said to be a *constant function*.

Exercise 1 (7–9 minutes)

Read through the opening text with students. Remind students that knowing the slope of the line that represents the function tells them if the function is increasing or decreasing. Introduce the term *constant function*. Present examples of functions that are constant; for example, your cell phone bill is $79 every month for unlimited calls and data. Let students work independently on Exercise 1; then, discuss and confirm answers as a class.

> **Exercises**
>
> 1. Read through each of the scenarios, and choose the graph of the function that best matches the situation. Explain the reason behind each choice.
>   a. A bathtub is filled at a constant rate of 1.75 gallons per minute.
>   b. A bathtub is drained at a constant rate of 2.5 gallons per minute.
>   c. A bathtub contains 2.5 gallons of water.
>   d. A bathtub is filled at a constant rate of 2.5 gallons per minute.

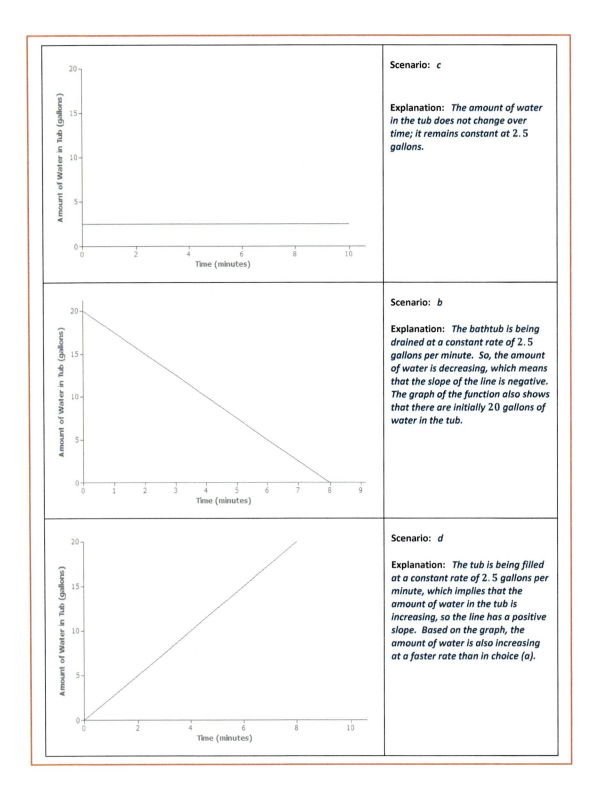

A STORY OF RATIOS   Lesson 4 8•6

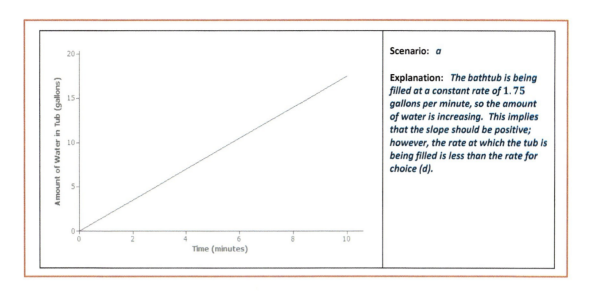

Scenario: *a*

Explanation: *The bathtub is being filled at a constant rate of 1.75 gallons per minute, so the amount of water is increasing. This implies that the slope should be positive; however, the rate at which the tub is being filled is less than the rate for choice (d).*

Exercise 2 (8–10 minutes)

In this exercise, students sketch a graph of a functional relationship given a verbal description. Allow students to work with a partner, and then confirm answers as a class. Refer to the functions as increasing or decreasing when discussing answers.

Students may misinterpret the meaning of *flat rate* in part (a). Discuss the meaning as a class. Tell students that it could also be called a *flat fee*.

After students have graphed the scenario presented in part (b), consider generating another graph where "meters under water" is represented using negative numbers. This provides an opportunity for students to see a real-world scenario with a negative slope graphed in the second quadrant. Ask students if both graphs model the same situation.

2. Read through each of the scenarios, and sketch a graph of a function that models the situation.

   a. A messenger service charges a flat rate of $4.95 to deliver a package regardless of the distance to the destination.

   The delivery charge remains constant regardless of the distance to the destination.

   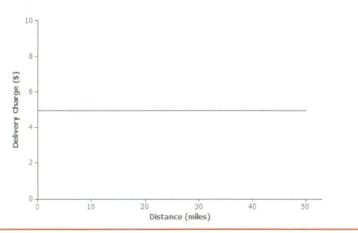

Lesson 4:   Increasing and Decreasing Functions   41

b. At sea level, the air that surrounds us presses down on our bodies at 14.7 pounds per square inch (psi). For every 10 meters that you dive under water, the pressure increases by 14.7 psi.

The initial value is 14.7 psi. The function increases at a rate of 14.7 psi for every 10 meters, or 1.47 psi per meter.

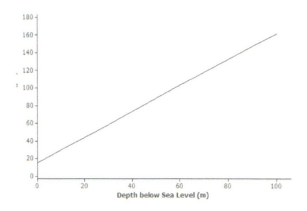

c. The range (driving distance per charge) of an electric car varies based on the average speed the car is driven. The initial range of the electric car after a full charge is 400 miles. However, the range is reduced by 20 miles for every 10 mph increase in average speed the car is driven.

The initial value of the function is 400. The function is decreasing by 20 miles for every 10 mph increase in speed. In other words, the function decreases by 2 miles for every 1 mph increase in speed.

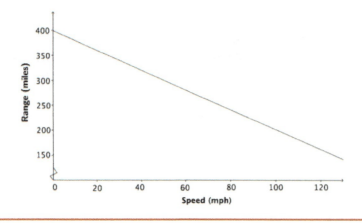

Exercise 3 (7–9 minutes)

Graphs of piecewise functions are introduced in this exercise. Students match verbal descriptions to a given graph. Let students work with a partner. Then, discuss and confirm answers as a class.

3. The graph below represents the total number of smartphones that are shipped to a retail store over the course of 50 days.

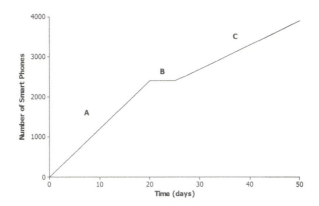

Match each part of the graph (A, B, and C) to its verbal description. Explain the reasoning behind your choice.

i. Half of the factory workers went on strike, and not enough smartphones were produced for normal shipments.

C; if half of the workers went on strike, then the number of smartphones produced would be less than normal. The rate of change for C is less than the rate of change for A.

ii. The production schedule was normal, and smartphones were shipped to the retail store at a constant rate.

A; if the production schedule is normal, the rate of change of interval A is greater than the rate of change of interval C.

iii. A defective electronic chip was found, and the factory had to shut down, so no smartphones were shipped.

B; if no smartphones are shipped to the store, the total number remains constant during that time.

Exercise 4 (10–12 minutes)

Let students work in small groups to create a story around the function represented by the graph. Then, compare stories as a class. Consider asking the following questions to connect the graph of the function to real-world experiences before groups begin writing their stories.

- What reason might explain why the account balance increases between Days 6 and 9 and then decreases between Days 9 and 14?
    - *Answers will vary. Maybe the person holding the account earned $15 each day mowing lawns and deposited the money each day to his account. Then, the same person needed to debit his account $6 each day to pay for lunch.*
- What reason might explain why the account balance does not change during the first few days?
    - *Answers will vary. Jameson is sick and cannot work to earn money to deposit into his account.*

Lesson 4: Increasing and Decreasing Functions

4. The relationship between Jameson's account balance and time is modeled by the graph below.

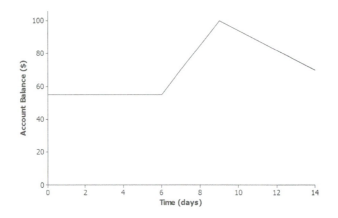

a. Write a story that models the situation represented by the graph.

Answers will vary.

Jameson was sick and did not work for almost a whole week. Then, he mowed several lawns over the next few days and deposited the money into his account after each job. It rained several days, so instead of working, Jameson withdrew money from his account each day to go to the movies and out to lunch with friends.

b. When is the function represented by the graph increasing? How does this relate to your story?

It is increasing between 6 and 9 days. Jameson earned money mowing lawns and made a deposit to his account each day. The money earned for each day was constant for these days. This is represented by a straight line.

c. When is the function represented by the graph decreasing? How does this relate to your story?

It is decreasing between 9 and 14 days. Since Days 9–14 are represented by a straight line, this means that Jameson spent the money constantly over these days. Jameson cannot work because it is raining. Perhaps he withdraws money from his account to spend on different activities each day because he cannot work.

Closing (3–4 minutes)

Review the Lesson Summary with the class.

- Refer back to Exercise 1. In parts (a) and (d), the bathtub was being filled at a constant rate. Is it reasonable within the context of the problem for the function in the graph to continue increasing?
  - *No. At some point, the tub will be full, and the amount of water cannot continue to increase.*
- Refer back to Exercise 2, part (b). The amount of pressure that an underwater diver experiences continues to increase as the diver continues to descend. Is it reasonable within the context of the problem for the function in the graph to continue increasing?
  - *No. At some point, the pressure will be too great, and the diver will not be able to descend any farther.*
- Is there a scenario that would require a function that modeled the situation to increase indefinitely? Explain.
  - *Yes. Students may use the example of money left in a savings account.*

It may need to be pointed out that this scenario is not necessarily linear, but if no money is withdrawn, the total will continue to increase.

> Lesson Summary
>
> The graph of a function can be used to help describe the relationship between two types of quantities.
>
> The slope of the line can provide useful information about the functional relationship between the quantities represented by the line:
>
> - A function whose graph has a positive slope is said to be an *increasing function*.
> - A function whose graph has a negative slope is said to be a *decreasing function*.
> - A function whose graph has a zero slope is said to be a *constant function*.

Exit Ticket (8 minutes)

Lesson 4: Increasing and Decreasing Functions

Exit Ticket

1. The graph below shows the relationship between a car's value and time.

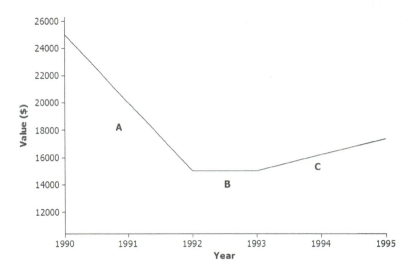

Match each part of the graph (A, B, and C) to its verbal description. Explain the reasoning behind your choice.

i. The value of the car holds steady due to a positive consumer report on the same model.

ii. There is a shortage of used cars on the market, and the value of the car rises at a constant rate.

iii. The value of the car depreciates at a constant rate.

2. Henry and Roxy both drive electric cars that need to be recharged before use. Henry uses a standard charger at his home to recharge his car. The graph below represents the relationship between the battery charge and the amount of time it has been connected to the power source for Henry's car.

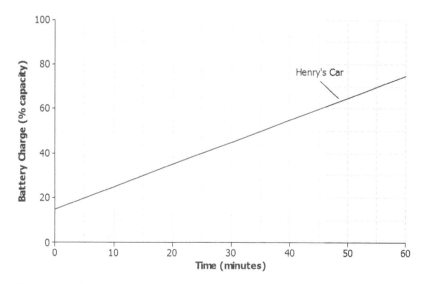

a. Describe how Henry's car battery is being recharged with respect to time.

b. Roxy has a supercharger at her home that can charge about half of the battery in 20 minutes. There is no remaining charge left when she begins recharging the battery. Sketch a graph that represents the relationship between the battery charge and the amount of time on the axes above. Assume the relationship is linear.

c. Which person's car will be recharged to full capacity first? Explain.

Exit Ticket Sample Solutions

1. The graph below shows the relationship between a car's value and time.

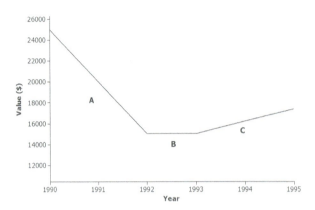

Match each part of the graph (A, B, and C) to its verbal description. Explain the reasoning behind your choice.

i. The value of the car holds steady due to a positive consumer report on the same model.

   B; if the value is holding steady, there is no change in the car's value between years.

ii. There is a shortage of used cars on the market, and the value of the car rises at a constant rate.

   C; if the value of the car is rising, it represents an increasing function.

iii. The value of the car depreciates at a constant rate.

   A; if the value depreciates, it represents a decreasing function.

2. Henry and Roxy both drive electric cars that need to be recharged before use. Henry uses a standard charger at his home to recharge his car. The graph below represents the relationship between the battery charge and the amount of time it has been connected to the power source for Henry's car.

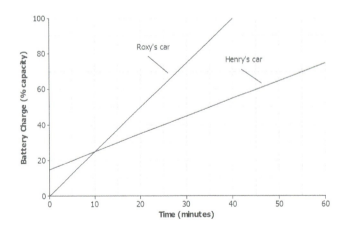

a. Describe how Henry's car battery is being recharged with respect to time.

   The battery charge is increasing at a constant rate of 10% every 10 minutes.

A STORY OF RATIOS   Lesson 4   8•6

b. Roxy has a supercharger at her home that can charge about half of the battery in 20 minutes. There is no remaining charge left when she begins recharging the battery. Sketch a graph that represents the relationship between the battery charge and the amount of time on the axes above. Assume the relationship is linear.

See the graph on the previous page.

c. Which person's car will be recharged to full capacity first? Explain.

Roxy's car will be completely recharged first. Her supercharger has a greater rate of change compared to Henry's charger.

Problem Set Sample Solutions

1. Read through each of the scenarios, and choose the graph of the function that best matches the situation. Explain the reason behind each choice.

    a. The tire pressure on Regina's car remains at 30 psi.
    b. Carlita inflates her tire at a constant rate for 4 minutes.
    c. Air is leaking from Courtney's tire at a constant rate.

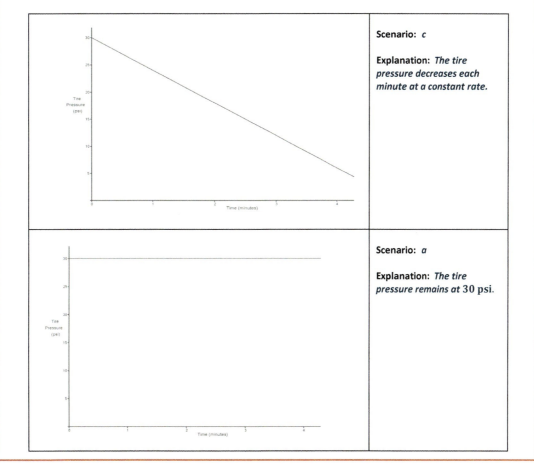

Scenario: *c*

Explanation: *The tire pressure decreases each minute at a constant rate.*

Scenario: *a*

Explanation: *The tire pressure remains at 30 psi.*

Lesson 4:   Increasing and Decreasing Functions

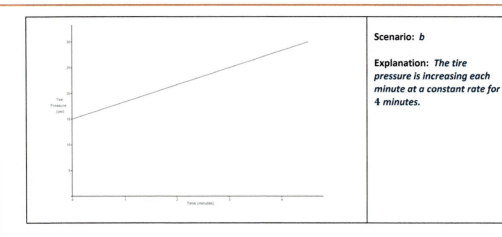

Scenario: *b*

Explanation: *The tire pressure is increasing each minute at a constant rate for 4 minutes.*

2. A home was purchased for $275,000. Due to a recession, the value of the home fell at a constant rate over the next 5 years.

    a. Sketch a graph of a function that models the situation.

    Graphs will vary; a sample graph is provided.

    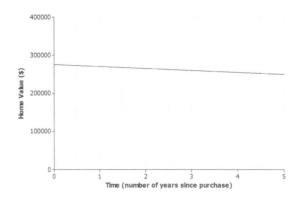

    b. Based on your graph, how is the home value changing with respect to time?

    Answers will vary; a sample answer is provided.

    The value is decreasing by $25,000 over 5 years or at a constant rate of $5,000 per year.

3. The graph below displays the first hour of Sam's bike ride.

    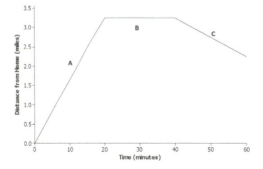

Lesson 4: Increasing and Decreasing Functions

Match each part of the graph (A, B, and C) to its verbal description. Explain the reasoning behind your choice.

i. Sam rides his bike to his friend's house at a constant rate.

A; the distance from home should be increasing as Sam is riding toward his friend's house.

ii. Sam and his friend bike together to an ice cream shop that is between their houses.

C; Sam was at his friend's house, but as they start biking to the ice cream shop, the distance from Sam's home begins to decrease.

iii. Sam plays at his friend's house.

B; Sam remains at the same distance from home while he is at his friend's house.

4. Using the axes below, create a story about the relationship between two quantities.

   a. Write a story about the relationship between two quantities. Any quantities can be used (e.g., distance and time, money and hours, age and growth). Be creative. Include keywords in your story such as *increase* and *decrease* to describe the relationship.

   Answers will vary. Give students the freedom to write a basic linear story or a piecewise story.

   A rock climber begins her descent from a height of 50 feet. She slowly descends at a constant rate for 4 minutes. She takes a break for 1 minute; she then realizes she left some of her gear on top of the rock and climbs more quickly back to the top at a constant rate.

   b. Label each axis with the quantities of your choice, and sketch a graph of the function that models the relationship described in the story.

   Answers will vary based on the story from part (a).

Lesson 4: Increasing and Decreasing Functions

Lesson 5: Increasing and Decreasing Functions

Student Outcomes

- Students qualitatively describe the functional relationship between two types of quantities by analyzing a graph.
- Students sketch a graph that exhibits the qualitative features of linear and nonlinear functions based on a verbal description.

Lesson Notes

This lesson extends the concepts introduced in Lesson 4 and focuses on graphs and the role they play in analyzing functional relationships between quantities. Students begin the lesson by comparing and contrasting linear and nonlinear functions. Encourage students to distinguish a linear function from a nonlinear function by analyzing a graph using the rate of change for an interval instead of just stating that "it looks like a straight line." Students sketch nonlinear functions given a contextual situation but do not construct the functions.

Classwork

Example 1 (3–5 minutes): Nonlinear Functions in the Real World

Read through the scenarios as a class. A linear function is used to model the first scenario, and a nonlinear function is used to model the second scenario.

> **Example 1: Nonlinear Functions in the Real World**
>
> Not all real-world situations can be modeled by a linear function. There are times when a nonlinear function is needed to describe the relationship between two types of quantities. Compare the two scenarios:
>
> a. Aleph is running at a constant rate on a flat, paved road. The graph below represents the total distance he covers with respect to time.
>
>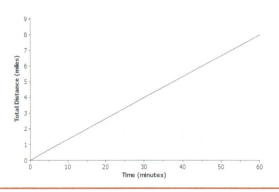

b. Shannon is running on a flat, rocky trail that eventually rises up a steep mountain. The graph below represents the total distance she covers with respect to time.

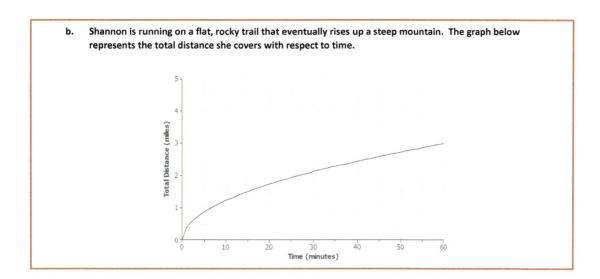

Exercises 1–2 (5–7 minutes)

Students look at the rate of change for different intervals for the scenarios presented in Example 1. Let students work with a partner. Then, discuss answers as a class. Remind students of increasing, decreasing, and constant *linear* functions from the previous lesson.

- Why might the distance that Shannon runs during each 15-minute interval decrease?
  - *Shannon is running up a mountain. Maybe the mountain is getting steeper, which is causing her to run slower.*
- Are these increasing or decreasing functions?
  - *They are both increasing functions because the total distance is increasing with respect to time. The function that models Aleph's total distance is an increasing linear function, and Shannon's total distance is an increasing nonlinear function.*

Exercises 1–2

1. In your own words, describe what is happening as Aleph is running during the following intervals of time.

   a. 0 to 15 minutes

   Aleph runs 2 miles in 15 minutes.

   b. 15 to 30 minutes

   Aleph runs another 2 miles in 15 minutes for a total of 4 miles.

   c. 30 to 45 minutes

   Aleph runs another 2 miles in 15 minutes for a total of 6 miles.

   d. 45 to 60 minutes

   Aleph runs another 2 miles in 15 minutes for a total of 8 miles.

2. In your own words, describe what is happening as Shannon is running during the following intervals of time.

   a. 0 to 15 minutes

   Shannon runs 1.5 miles in 15 minutes.

   b. 15 to 30 minutes

   Shannon runs another 0.6 mile in 15 minutes for a total of 2.1 miles.

   c. 30 to 45 minutes

   Shannon runs another 0.5 mile in 15 minutes for a total of 2.6 miles.

   d. 45 to 60 minutes

   Shannon runs another 0.4 mile in 15 minutes for a total of 3.0 miles.

Example 2 (5 minutes): Increasing and Decreasing Functions

Convey to students that linear functions have a constant rate of change while nonlinear functions *do not* have a constant rate of change. Consider using a table of values for additional clarification using the information from Exercises 1 and 2.

- How would you describe the rate of change of the function modeling Shannon's total distance? Explain.
  - *The function is increasing but at a decreasing rate of change. The rate of change is decreasing for every 15-minute interval.*

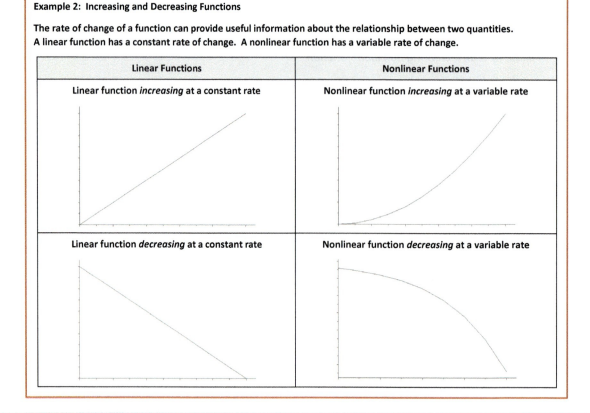

Lesson 5: Increasing and Decreasing Functions

A STORY OF RATIOS   Lesson 5   8•6

Linear function with a constant rate	Nonlinear function with a variable rate
x : y 0 : 7 1 : 10 2 : 13 3 : 16 4 : 19	x : y 0 : 0 1 : 2 2 : 4 3 : 8 4 : 16

Exercises 3–5 (15 minutes)

Students sketch graphs of functions based on a verbal description. Note that the graph should just be a rough sketch that matches the verbal description. Allow students to work with a partner or in a small group. Discuss and compare answers as a class.

> **Exercises 3–5**
>
> 3. Different breeds of dogs have different growth rates. A large breed dog typically experiences a rapid growth rate from birth to age 6 months. At that point, the growth rate begins to slow down until the dog reaches full growth around 2 years of age.
>
>   a. Sketch a graph that represents the weight of a large breed dog from birth to 2 years of age.
>
>   *Answers will vary.*
>
>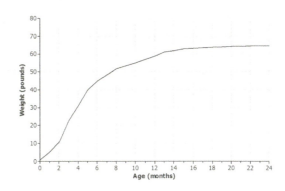
>
>   b. Is the function represented by the graph linear or nonlinear? Explain.
>
>   *The function is nonlinear because the growth rate is not constant.*
>
>   c. Is the function represented by the graph increasing or decreasing? Explain.
>
>   *The function is increasing but at a decreasing rate. There is rapid growth during the first 6 months, and then the growth rate decreases.*

Lesson 5:   Increasing and Decreasing Functions

4. Nikka took her laptop to school and drained the battery while typing a research paper. When she returned home, Nikka connected her laptop to a power source, and the battery recharged at a constant rate.

   a. Sketch a graph that represents the battery charge with respect to time.

   Answers will vary.

   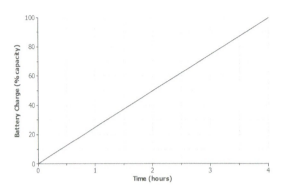

   b. Is the function represented by the graph linear or nonlinear? Explain.

   The function is linear because the battery is recharging at a constant rate.

   c. Is the function represented by the graph increasing or decreasing? Explain.

   The function is increasing because the battery is being recharged.

5. The long jump is a track-and-field event where an athlete attempts to leap as far as possible from a given point. Mike Powell of the United States set the long jump world record of 8.95 meters (29.4 feet) during the 1991 World Championships in Tokyo, Japan.

   a. Sketch a graph that represents the path of a high school athlete attempting the long jump.

   Answers will vary.

   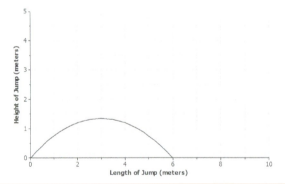

Note: If students have trouble visualizing the path of a jump, use the following table for students to begin their sketches. Remind students to draw a curve and not to connect points with a straight line.

x	y
0	0
1	0.75
2	1.2
3	1.35
4	1.2
5	0.75
6	0

> b. Is the function represented by the graph linear or nonlinear? Explain.
>
> *The function is nonlinear. The rate of change is not constant.*
>
> c. Is the function represented by the graph increasing or decreasing? Explain.
>
> *The function both increases and decreases over different intervals. The function increases as the athlete begins the jump and reaches a maximum height. The function decreases after the athlete reaches maximum height and begins descending back toward the ground.*

Example 3 (5–7 minutes): Ferris Wheel

This example presents students with a graph of a nonlinear function that both increases and decreases over different intervals of time. Students may have a difficult time connecting the graph to the scenario. Remind students that the graph is relating time to a rider's distance above the ground. Consider doing a rough sketch of the Ferris wheel scenario on a personal white board for further clarification using a similar object such as a hamster wheel or a K'NEX construction toy. There are also videos that can be found online that relate this type of motion to nonlinear curves. The website www.graphingstories.com has a great video that relates the motion of a playground merry-go-round to the distance of a camera that produces a graph similar to the Ferris wheel example.

> **Example 3: Ferris Wheel**
>
> Lamar and his sister are riding a Ferris wheel at a state fair. Using their watches, they find that it takes 8 seconds for the Ferris wheel to make a complete revolution. The graph below represents Lamar and his sister's distance above the ground with respect to time.
>
>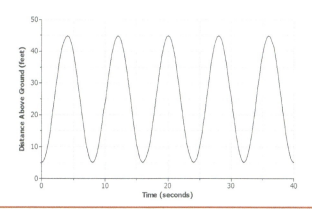

Lesson 5: Increasing and Decreasing Functions

A STORY OF RATIOS   Lesson 5   8•6

Exercises 6–9 (5–7 minutes)

Allow students to work with a partner or in a small group to complete the following exercises. Confirm answers as a class.

Exercises 6–9

6. Use the graph from Example 3 to answer the following questions.

   a. Is the function represented by the graph linear or nonlinear?

   The function is nonlinear. The rate of change is not constant.

   b. Where is the function increasing? What does this mean within the context of the problem?

   The function is increasing during the following intervals of time: 0 to 4 seconds, 8 to 12 seconds, 16 to 20 seconds, 24 to 28 seconds, and 32 to 36 seconds. It means that Lamar and his sister are rising in the air.

   c. Where is the function decreasing? What does this mean within the context of the problem?

   The function is decreasing during the following intervals of time: 4 to 8 seconds, 12 to 16 seconds, 20 to 24 seconds, 28 to 32 seconds, and 36 to 40 seconds. Lamar and his sister are traveling back down toward the ground.

7. How high above the ground is the platform for passengers to get on the Ferris wheel? Explain your reasoning.

   The lowest point on the graph, which is at 5 feet, can represent the platform where the riders get on the Ferris wheel.

8. Based on the graph, how many revolutions does the Ferris wheel complete during the 40-second time interval? Explain your reasoning.

   The Ferris wheel completes 5 revolutions. The lowest points on the graph can represent Lamar and his sister at the beginning of a revolution or at the entrance platform of the Ferris wheel. So, one revolution occurs between 0 and 8 seconds, 8 and 16 seconds, 16 and 24 seconds, 24 and 32 seconds, and 32 and 40 seconds.

9. What is the diameter of the Ferris wheel? Explain your reasoning.

   The diameter of the Ferris wheel is 40 feet. The lowest point on the graph represents the base of the Ferris wheel, and the highest point on the graph represents the top of the Ferris wheel. The difference between the two values is 40 feet, which is the diameter of the wheel.

Closing (2 minutes)

Review the Lesson Summary with students.

- Refer back to Exercises 3 and 4 (dog growth rate and laptop battery recharge problems). Note that both functions were increasing. Is it possible for those functions to continue to increase within the context of the problem? Explain.
    - *No. Both functions cannot continue to increase.*
    - *The dog's weight will increase until it reaches full growth. At that point, the weight would remain constant or may fluctuate based on diet and exercise.*
    - *The laptop battery capacity can only reach 100%. At that point, it is fully charged. The function could not continue to increase.*

Lesson 5: Increasing and Decreasing Functions

> **Lesson Summary**
>
> The graph of a function can be used to help describe the relationship between the quantities it represents.
>
> A linear function has a constant rate of change. A nonlinear function does not have a constant rate of change.
>
> - A function whose graph has a positive rate of change is an *increasing function*.
> - A function whose graph has a negative rate of change is a *decreasing function*.
> - Some functions may increase and decrease over different intervals.

Exit Ticket (5 minutes)

Name _____ Date _____

Lesson 5: Increasing and Decreasing Functions

Exit Ticket

Lamar and his sister continue to ride the Ferris wheel. The graph below represents Lamar and his sister's distance above the ground with respect to time during the next 40 seconds of their ride.

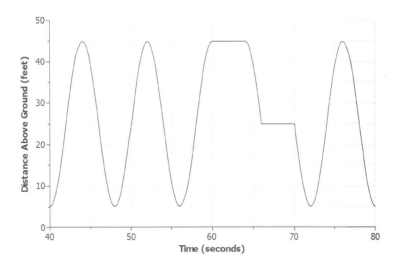

a. Name one interval where the function is increasing.

b. Name one interval where the function is decreasing.

c. Is the function linear or nonlinear? Explain.

d. What could be happening during the interval of time from 60 to 64 seconds?

e. Based on the graph, how many complete revolutions are made during this 40-second interval?

Exit Ticket Sample Solutions

Lamar and his sister continue to ride the Ferris wheel. The graph below represents Lamar and his sister's distance above the ground with respect to time during the next 40 seconds of their ride.

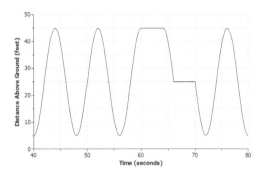

a. Name one interval where the function is increasing.

The function is increasing during the following intervals of time: 40 to 44 seconds, 48 to 52 seconds, 56 to 60 seconds, and 72 to 76 seconds.

b. Name one interval where the function is decreasing.

The function is decreasing during the following intervals of time: 44 to 48 seconds, 52 to 56 seconds, 64 to 66 seconds, 70 to 72 seconds, and 76 to 80 seconds.

c. Is the function linear or nonlinear? Explain.

The function is both linear and nonlinear during different intervals of time. It is linear from 60 to 64 seconds and from 66 to 70 seconds. It is nonlinear from 40 to 60 seconds and from 70 to 80 seconds.

d. What could be happening during the interval of time from 60 to 64 seconds?

The Ferris wheel is not moving during that time, so riders may be getting off or getting on.

e. Based on the graph, how many complete revolutions are made during this 40-second interval?

Four revolutions are made during this time period.

Problem Set Sample Solutions

1. Read through the following scenarios, and match each to its graph. Explain the reasoning behind your choice.

   a. This shows the change in a smartphone battery charge as a person uses the phone more frequently.
   b. A child takes a ride on a swing.
   c. A savings account earns simple interest at a constant rate.
   d. A baseball has been hit at a youth baseball game.

   Scenario: *c*

   The savings account is earning interest at a constant rate, which means that the function is linear.

   Scenario: *d*

   The baseball is hit into the air, reaches a maximum height, and falls back to the ground at a variable rate.

   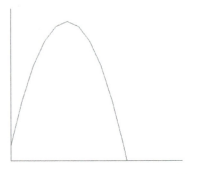

   Scenario: *b*

   The distance from the ground increases as the child swings up into the air and then decreases as the child swings back down toward the ground.

   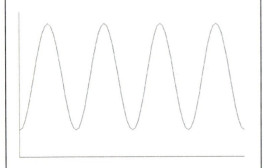

   Scenario: *a*

   The battery charge is decreasing as a person uses the phone more frequently.

   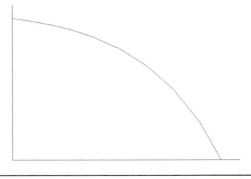

Lesson 5: Increasing and Decreasing Functions

2. The graph below shows the volume of water for a given creek bed during a 24-hour period. On this particular day, there was wet weather with a period of heavy rain.

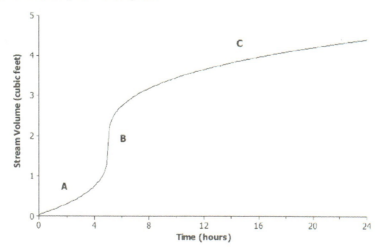

Describe how each part (A, B, and C) of the graph relates to the scenario.

A: The rain begins, and the volume of water flowing in the creek bed begins to increase.

B: A period of heavy rain occurs, causing the volume of water to increase.

C: The heavy rain begins to subside, and the volume of water continues to increase.

3. Half-life is the time required for a quantity to fall to half of its value measured at the beginning of the time period. If there are 100 grams of a radioactive element to begin with, there will be 50 grams after the first half-life, 25 grams after the second half-life, and so on.

   a. Sketch a graph that represents the amount of the radioactive element left with respect to the number of half-lives that have passed.

   Answers will vary.

   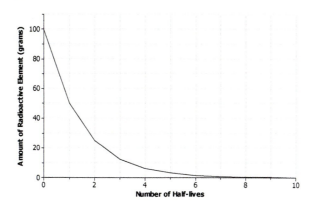

   b. Is the function represented by the graph linear or nonlinear? Explain.

   The function is nonlinear. The rate of change is not constant with respect to time.

   c. Is the function represented by the graph increasing or decreasing?

   The function is decreasing.

4. Lanae parked her car in a no-parking zone. Consequently, her car was towed to an impound lot. In order to release her car, she needs to pay the impound lot charges. There is an initial charge on the day the car is brought to the lot. However, 10% of the previous day's charges will be added to the total charge for every day the car remains in the lot.

   a. Sketch a graph that represents the total charges with respect to the number of days a car remains in the impound lot.

   Answers will vary.

   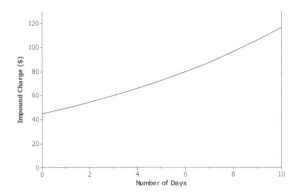

   b. Is the function represented by the graph linear or nonlinear? Explain.

   The function is nonlinear. The function is increasing.

   c. Is the function represented by the graph increasing or decreasing? Explain.

   The function is increasing. The total charge is increasing as the number of days the car is left in the lot increases.

5. Kern won a $50 gift card to his favorite coffee shop. Every time he visits the shop, he purchases the same coffee drink.

   a. Sketch a graph of a function that can be used to represent the amount of money that remains on the gift card with respect to the number of drinks purchased.

   Answers will vary.

   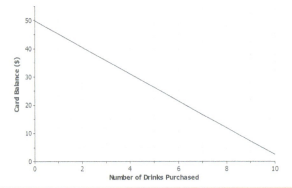

Lesson 5: Increasing and Decreasing Functions

b. Is the function represented by the graph linear or nonlinear? Explain.

The function is linear. Since Kern purchases the same drink every visit, the balance is decreasing by the same amount or, in other words, at a constant rate of change.

c. Is the function represented by the graph increasing or decreasing? Explain.

The function is decreasing. With each drink purchased, the amount of money on the card decreases.

6. Jay and Brooke are racing on bikes to a park 8 miles away. The tables below display the total distance each person biked with respect to time.

Jay

Time (minutes)	Distance (miles)
0	0
5	0.84
10	1.86
15	3.00
20	4.27
25	5.67

Brooke

Time (minutes)	Distance (miles)
0	0
5	1.2
10	2.4
15	3.6
20	4.8
25	6.0

a. Which person's biking distance could be modeled by a nonlinear function? Explain.

The distance that Jay biked could be modeled by a nonlinear function because the rate of change is not constant. The distance that Brooke biked could be modeled by a linear function because the rate of change is constant.

b. Who would you expect to win the race? Explain.

Jay will win the race. The distance he bikes during each five-minute interval is increasing, while Brooke's biking distance remains constant. If the trend remains the same, it is estimated that both Jay and Brooke will travel about 7.2 miles in 30 minutes. So, Jay will overtake Brooke during the last 5 minutes to win the race.

7. Using the axes in Problem 7(b), create a story about the relationship between two quantities.

a. Write a story about the relationship between two quantities. Any quantities can be used (e.g., distance and time, money and hours, age and growth). Be creative! Include keywords in your story such as *increase* and *decrease* to describe the relationship.

Answers will vary.

A person in a car is at a red stoplight. The light turns green, and the person presses down on the accelerator with increasing pressure. The car begins to move and accelerate. The rate at which the car accelerates is not constant.

Lesson 5: Increasing and Decreasing Functions

b. Label each axis with the quantities of your choice, and sketch a graph of the function that models the relationship described in the story.

Answers will vary based on the story from part (a).

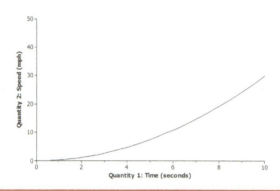

A STORY OF RATIOS

Mathematics Curriculum

GRADE 8 • MODULE 6

Topic B
Bivariate Numerical Data

Focus Standards:	▪ Construct and interpret scatter plots for bivariate measurement data to investigate patterns of association between two quantities. Describe patterns such as clustering, outliers, positive or negative association, linear association, and nonlinear association.
	▪ Know that straight lines are widely used to model relationships between two quantitative variables. For scatter plots that suggest a linear association, informally fit a straight line, and informally assess the model fit by judging the closeness of the data points to the line.
Instructional Days:	4
Lesson 6:	Scatter Plots (P)[1]
Lesson 7:	Patterns in Scatter Plots (P)
Lesson 8:	Informally Fitting a Line (P)
Lesson 9:	Determining the Equation of a Line Fit to Data (P)

In Topic B, students connect their study of linear functions to applications involving bivariate data. A key tool in developing this connection is a scatter plot. In Lesson 6, students construct scatter plots and focus on identifying linear versus nonlinear patterns. They distinguish positive linear association and negative linear association based on the scatter plot. Students describe trends in the scatter plot along with clusters and outliers (points that do not fit the pattern). In Lesson 8, students informally fit a straight line to data displayed in a scatter plot by judging the closeness of the data points to the line. In Lesson 9, students interpret and determine the equation of the line they fit to the data and use the equation to make predictions and to evaluate possible association of the variables. Based on these predictions, students address the need for a *best-fit* line, which is formally introduced in Algebra I.

---

[1]Lesson Structure Key: **P**-Problem Set Lesson, **M**-Modeling Cycle Lesson, **E**-Exploration Lesson, **S**-Socratic Lesson

Topic B: Bivariate Numerical Data

Lesson 6: Scatter Plots

Student Outcomes

- Students construct scatter plots.
- Students use scatter plots to investigate relationships.
- Students understand that a trend in a scatter plot does not establish cause-and-effect.

Lesson Notes

This lesson is the first in a set of lessons dealing with relationships between numerical variables. In this lesson, students learn how to construct a scatter plot and look for patterns that suggest a statistical relationship between two numerical variables. Note that in this, and subsequent lessons, there is notation on the graphs indicating that not all of the intervals are represented on the axes. Students may need explanation that connects the "zigzag" to the idea that there are numbers on the axes that are just not shown in the graph.

Classwork

Example 1 (5 minutes)

Spend a few minutes introducing the context of this example. Make sure that students understand that in this context, an *observation* can be thought of as an ordered pair consisting of the value for each of two variables.

Scaffolding:

- Point out to students that the word *bivariate* is composed of the prefix *bi-* and the stem *variate*.
- *Bi-* means *two*.
- *Variate* indicates a variable.
- The focus in this lesson is on two numerical variables.

Example 1

A bivariate data set consists of observations on two variables. For example, you might collect data on 13 different car models. Each observation in the data set would consist of an (x, y) pair.

x: weight (in pounds, rounded to the nearest 50 pounds)

and

y: fuel efficiency (in miles per gallon, mpg)

The table below shows the weight and fuel efficiency for 13 car models with automatic transmissions manufactured in 2009 by Chevrolet.

Model	Weight (pounds)	Fuel Efficiency (mpg)
1	3,200	23
2	2,550	28
3	4,050	19
4	4,050	20
5	3,750	20
6	3,550	22
7	3,550	19
8	3,500	25
9	4,600	16
10	5,250	12
11	5,600	16
12	4,500	16
13	4,800	15

Scaffolding:

- English language learners new to the curriculum may be familiar with the metric system (kilometers, kilograms, and liters) but unfamiliar with the English system (miles, pounds, and gallons).
- It may be helpful to provide conversions:
  $1 \text{ kg} \approx 2.2 \text{ lb.}$
  $1 \text{ lb.} \approx 0.45 \text{ kg}$
  $1 \text{ km} \approx 0.62 \text{ mi.}$
  $1 \text{ mi.} \approx 1.61 \text{ km}$

Exercises 1–3 (10–12 minutes)

After students have had a chance to think about Exercise 1, make sure that everyone understands what an observation (an ordered pair) represents in the context of this example. Relate plotting the point that corresponds to the first observation to students' previous work with plotting points in a rectangular coordinate system. As a way of encouraging the need to look at a graph of the data, consider asking students to try to determine if there is a relationship between weight and fuel efficiency by just looking at the table. Allow students time to complete the scatter plot and complete Exercise 3. Have students share their answers to Exercise 3.

Exercises 1–8

1. In the Example 1 table, the observation corresponding to Model 1 is $(3200, 23)$. What is the fuel efficiency of this car? What is the weight of this car?

   The fuel efficiency is 23 miles per gallon, and the weight is 3,200 pounds.

One question of interest is whether there is a relationship between the car weight and fuel efficiency. The best way to begin to investigate is to construct a graph of the data. A *scatter plot* is a graph of the (x, y) pairs in the data set. Each (x, y) pair is plotted as a point in a rectangular coordinate system.

For example, the observation $(3200, 23)$ would be plotted as a point located above 3,200 on the x-axis and across from 23 on the y-axis, as shown below.

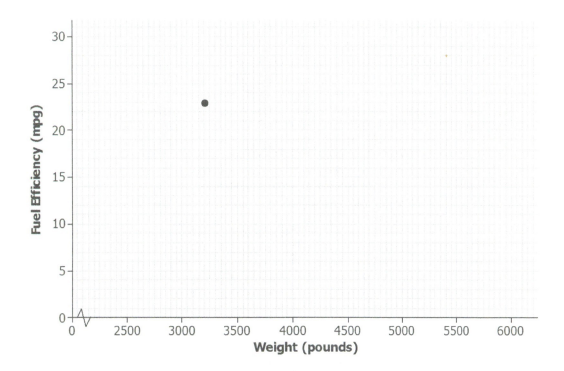

Lesson 6: Scatter Plots

2. Add the points corresponding to the other 12 observations to the scatter plot.

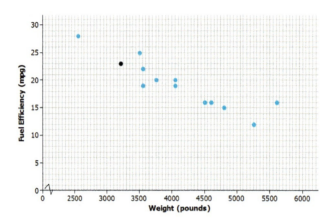

3. Do you notice a pattern in the scatter plot? What does this imply about the relationship between weight (x) and fuel efficiency (y)?

   There does seem to be a pattern in the plot. Higher weights tend to be paired with lesser fuel efficiencies, so it looks like heavier cars generally have lower fuel efficiency.

Exercises 4–8 (6–8 minutes)

These exercises give students additional practice creating a scatter plot and identifying a pattern in the plot. Students should work individually on these exercises and then discuss their answers to Exercises 7 and 8 with a partner. However, some English language learners may benefit from paired or small group work, particularly if their English literacy is not strong.

Is there a relationship between price and the quality of athletic shoes? The data in the table below are from the *Consumer Reports* website.

x: price (in dollars)

and

y: *Consumer Reports* quality rating

The quality rating is on a scale of 0 to 100, with 100 being the highest quality.

Shoe	Price (dollars)	Quality Rating
1	65	71
2	45	70
3	45	62
4	80	59
5	110	58
6	110	57
7	30	56
8	80	52
9	110	51
10	70	51

A STORY OF RATIOS   Lesson 6   8•6

4. One observation in the data set is $(110, 57)$. What does this ordered pair represent in terms of cost and quality?

   The pair represents a shoe that costs $110 with a quality rating of 57.

5. To construct a scatter plot of these data, you need to start by thinking about appropriate scales for the axes of the scatter plot. The prices in the data set range from $30 to $110, so one reasonable choice for the scale of the x-axis would range from $20 to $120, as shown below. What would be a reasonable choice for a scale for the y-axis?

   Sample response: The smallest y-value is 51, and the largest y-value is 71. So, the y-axis could be scaled from 50 to 75.

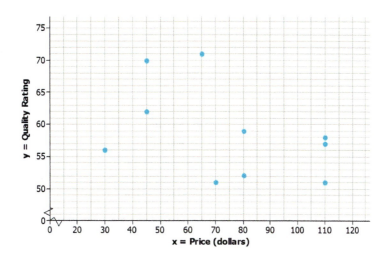

6. Add a scale to the y-axis. Then, use these axes to construct a scatter plot of the data.

7. Do you see any pattern in the scatter plot indicating that there is a relationship between price and quality rating for athletic shoes?

   Answers will vary. Students may say that they do not see a pattern, or they may say that they see a slight downward trend.

8. Some people think that if shoes have a high price, they must be of high quality. How would you respond?

   Answers will vary. The data do not support this. Students will either respond that there does not appear to be a relationship between price and quality, or if they saw a downward trend in the scatter plot, they might even indicate that the higher-priced shoes tend to have lower quality. Look for consistency between the answer to this question and how students answered the previous question.

Scaffolding:

For more complicated and reflective answers, consider allowing English language learners to use one or more of the following options: collaborate with a same-language peer, illustrate their responses, or provide a first-language narration or response.

Example 2 (5–10 minutes): Statistical Relationships

This example makes a very important point. While discussing this example with the class, make sure students understand the distinction between a statistical relationship and a cause-and-effect relationship. After discussing the example, ask students if they can think of other examples of numerical variables that might have a statistical relationship but which probably do not have a cause-and-effect relationship.

Lesson 6:   Scatter Plots   71

A STORY OF RATIOS   Lesson 6 8•6

> **Example 2: Statistical Relationships**
>
> A pattern in a scatter plot indicates that the values of one variable tend to vary in a predictable way as the values of the other variable change. This is called a *statistical relationship*. In the fuel efficiency and car weight example, fuel efficiency tended to decrease as car weight increased.
>
> This is useful information, but be careful not to jump to the conclusion that increasing the weight of a car *causes* the fuel efficiency to go down. There may be some other explanation for this. For example, heavier cars may also have bigger engines, and bigger engines may be less efficient. You cannot conclude that changes to one variable *cause* changes in the other variable just because there is a statistical relationship in a scatter plot.

Exercises 9–10 (5 minutes)

Students can work individually or with a partner on these exercises. Then, confirm answers as a class.

> **Exercises 9–10**
>
> 9. Data were collected on
>
> x: shoe size
>
>   and
>
> y: score on a reading ability test
>
>   for 29 elementary school students. The scatter plot of these data is shown below. Does there appear to be a statistical relationship between shoe size and score on the reading test?
>
>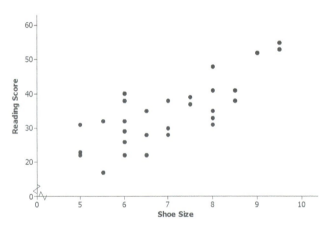
>
> *Possible response: The pattern in the scatter plot appears to follow a line. As shoe sizes increase, the reading scores also seem to increase. There does appear to be a statistical relationship because there is a pattern in the scatter plot.*
>
> 10. Explain why it is not reasonable to conclude that having big feet causes a high reading score. Can you think of a different explanation for why you might see a pattern like this?
>
>   *Possible response: You cannot conclude that just because there is a statistical relationship between shoe size and reading score that one causes the other. These data were for students completing a reading test for younger elementary school children. Older children, who would have bigger feet than younger children, would probably tend to score higher on a reading test for younger students.*

Lesson 6:   Scatter Plots

Closing (3 minutes)

Consider posing the following questions; allow a few student responses for each.

- Why is it helpful to make a scatter plot when you have data on two numerical variables?
  - *A scatter plot makes it easier to see patterns in the data and to see if there is a statistical relationship between the two variables.*
- Can you think of an example of two variables that would have a statistical relationship but not a cause-and-effect relationship?
  - *One famous example is the number of people who must be rescued by lifeguards at the beach and the number of ice cream sales. Both of these variables have higher values when the temperature is high and lower values when the temperature is low. So, there is a statistical relationship between them—they tend to vary in a predictable way. However, it would be silly to say that an increase in ice cream sales causes more beach rescues.*

Lesson Summary

- A scatter plot is a graph of numerical data on two variables.
- A pattern in a scatter plot suggests that there may be a relationship between the two variables used to construct the scatter plot.
- If two variables tend to vary together in a predictable way, we can say that there is a statistical relationship between the two variables.
- A statistical relationship between two variables does not imply that a change in one variable causes a change in the other variable (a cause-and-effect relationship).

Exit Ticket (5 minutes)

A STORY OF RATIOS   Lesson 6   8•6

Name _____   Date _____

Lesson 6: Scatter Plots

Exit Ticket

Energy is measured in kilowatt-hours. The table below shows the cost of building a facility to produce energy and the ongoing cost of operating the facility for five different types of energy.

Type of Energy	Cost to Operate (cents per kilowatt-hour)	Cost to Build (dollars per kilowatt-hour)
Hydroelectric	0.4	2,200
Wind	1.0	1,900
Nuclear	2.0	3,500
Coal	2.2	2,500
Natural Gas	4.8	1,000

1. Construct a scatter plot of the cost to build the facility in dollars per kilowatt-hour (x) and the cost to operate the facility in cents per kilowatt-hour (y). Use the grid below, and be sure to add an appropriate scale to the axes.

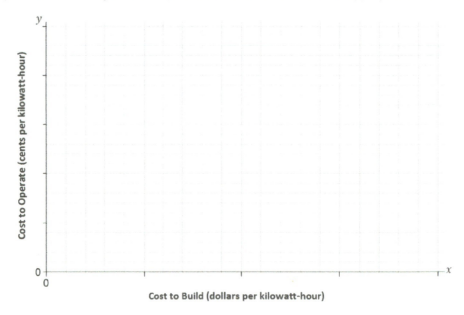

2. Do you think that there is a statistical relationship between building cost and operating cost? If so, describe the nature of the relationship.

3. Based on the scatter plot, can you conclude that decreased building cost is the cause of increased operating cost? Explain.

Exit Ticket Sample Solutions

Energy is measured in kilowatt-hours. The table below shows the cost of building a facility to produce energy and the ongoing cost of operating the facility for five different types of energy.

Type of Energy	Cost to Operate (cents per kilowatt-hour)	Cost to Build (dollars per kilowatt-hour)
Hydroelectric	0.4	2,200
Wind	1.0	1,900
Nuclear	2.0	3,500
Coal	2.2	2,500
Natural Gas	4.8	1,000

1. Construct a scatter plot of the cost to build the facility in dollars per kilowatt-hour (x) and the cost to operate the facility in cents per kilowatt-hour (y). Use the grid below, and be sure to add an appropriate scale to the axes.

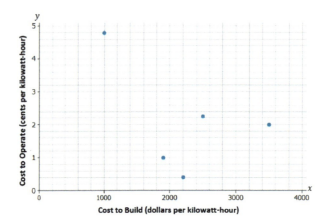

2. Do you think that there is a statistical relationship between building cost and operating cost? If so, describe the nature of the relationship.

   Answers may vary. Sample response: Yes, because it looks like there is a downward pattern in the scatter plot. It appears that the types of energy that have facilities that are more expensive to build are less expensive to operate.

3. Based on the scatter plot, can you conclude that decreased building cost is the cause of increased operating cost? Explain.

   Sample response: No. Just because there may be a statistical relationship between cost to build and cost to operate does not mean that there is a cause-and-effect relationship.

Problem Set Sample Solutions

The Problem Set is intended to reinforce material from the lesson and have students think about the meaning of points in a scatter plot, clusters, positive and negative linear trends, and trends that are not linear.

1. The table below shows the price and overall quality rating for 15 different brands of bike helmets.

   Data source: www.consumerreports.org

Helmet	Price (dollars)	Quality Rating
A	35	65
B	20	61
C	30	60
D	40	55
E	50	54
F	23	47
G	30	47
H	18	43
I	40	42
J	28	41
K	20	40
L	25	32
M	30	63
N	30	63
O	40	53

   Construct a scatter plot of price (x) and quality rating (y). Use the grid below.

   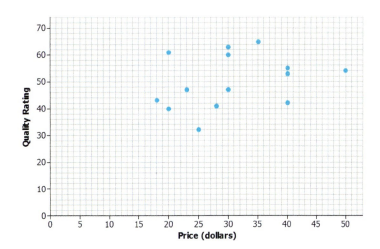

2. Do you think that there is a statistical relationship between price and quality rating? If so, describe the nature of the relationship.

   Sample response: No. There is no pattern visible in the scatter plot. There does not appear to be a relationship between price and the quality rating for bike helmets.

Lesson 6: Scatter Plots

3. Scientists are interested in finding out how different species adapt to finding food sources. One group studied crocodilian species to find out how their bite force was related to body mass and diet. The table below displays the information they collected on body mass (in pounds) and bite force (in pounds).

Species	Body Mass (pounds)	Bite Force (pounds)
Dwarf crocodile	35	450
Crocodile F	40	260
Alligator A	30	250
Caiman A	28	230
Caiman B	37	240
Caiman C	45	255
Crocodile A	110	550
Nile crocodile	275	650
Crocodile B	130	500
Crocodile C	135	600
Crocodile D	135	750
Caiman D	125	550
Indian Gharial crocodile	225	400
Crocodile G	220	1,000
American crocodile	270	900
Crocodile E	285	750
Crocodile F	425	1,650
American alligator	300	1,150
Alligator B	325	1,200
Alligator C	365	1,450

Data Source: http://journals.plos.org/plosone/article?id=10.1371/journal.pone.0031781#pone-0031781-t001

(Note: Body mass and bite force have been converted to pounds from kilograms and newtons, respectively.)

Construct a scatter plot of body mass (x) and bite force (y). Use the grid below, and be sure to add an appropriate scale to the axes.

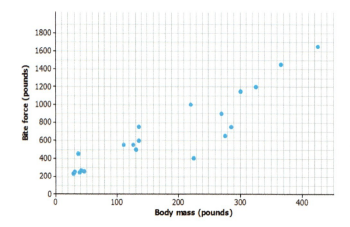

4. Do you think that there is a statistical relationship between body mass and bite force? If so, describe the nature of the relationship.

   Sample response: Yes, because it looks like there is an upward pattern in the scatter plot. It appears that alligators with larger body mass also tend to have greater bite force.

5. Based on the scatter plot, can you conclude that increased body mass causes increased bite force? Explain.

   Sample response: No. Just because there is a statistical relationship between body mass and bite force does not mean that there is a cause-and-effect relationship.

A STORY OF RATIOS   Lesson 7   8•6

Lesson 7: Patterns in Scatter Plots

Student Outcomes

- Students distinguish linear patterns from nonlinear patterns based on scatter plots.
- Students describe positive and negative trends in a scatter plot.
- Students identify and describe unusual features in scatter plots, such as clusters and outliers.

Lesson Notes

This lesson asks students to look for and describe patterns in scatter plots. It provides a foundation for later lessons in which students use a line to describe the relationship between two numerical variables when the pattern in the scatter plot is linear. Students distinguish between linear and nonlinear relationships as well as positive and negative linear relationships. The terms *clusters* and *outliers* are also introduced, and students look for these features in scatter plots and investigate what clusters and outliers reveal about the data.

> *Scaffolding:*
> - Point out to students that in this lesson, the meaning of the word *relationship* is not the same as the use of the word describing a familial connection, such as a sister or cousin.
> - In this lesson, a *relationship* indicates that two numerical variables have a connection that can be described either verbally or with mathematical symbols.

Classwork

Example 1 (3–5 minutes)

Spend a few minutes going over the three questions posed as a way to help students structure their thinking about data displayed in a scatter plot. Students should see that looking for patterns in a scatter plot is a logical extension of their work in the previous lesson where they learned to make a scatter plot. Make sure that students understand the distinction between a positive linear relationship and a negative linear relationship before moving on to Exercises 1–5. Students have a chance to practice answering these questions in the exercises that follow. Consider asking students to examine the five scatter plots and describe their similarities and differences before telling students what to look for.

> *Scaffolding:*
> For English language learners, teachers may need to read aloud the information in Example 1, highlighting each key point with a visual example as students record it in a graphic organizer for reference.

Example 1

In the previous lesson, you learned that scatter plots show trends in bivariate data.

When you look at a scatter plot, you should ask yourself the following questions:

    a. Does it look like there is a relationship between the two variables used to make the scatter plot?

    b. If there is a relationship, does it appear to be linear?

    c. If the relationship appears to be linear, is the relationship a positive linear relationship or a negative linear relationship?

To answer the first question, look for patterns in the scatter plot. Does there appear to be a general pattern to the points in the scatter plot, or do the points look as if they are scattered at random? If you see a pattern, you can answer the second question by thinking about whether the pattern would be well described by a line. Answering the third question requires you to distinguish between a positive linear relationship and a negative linear relationship. A positive linear relationship is one that is described by a line with a positive slope. A negative linear relationship is one that is described by a line with a negative slope.

Lesson 7:   Patterns in Scatter Plots

©2018 Great Minds®. eureka-math.org

Exercises 1–5 (8–10 minutes)

Consider answering Exercise 1 as part of a whole-class discussion, and then allow students to work individually or in pairs on Exercises 2–5. Have students share answers to these exercises, and discuss any of the exercises where there is disagreement on the answers. Additionally, point out to students that scatter plots that more closely resemble a linear pattern are sometimes called *strong*. Scatter plots that are linear but not as close to a line are sometimes known as *weak*. A linear relationship may sometimes be referred to as *strong positive*, *weak positive*, *strong negative*, or *weak negative*. Consider using these terms with students while discussing their scatter plots.

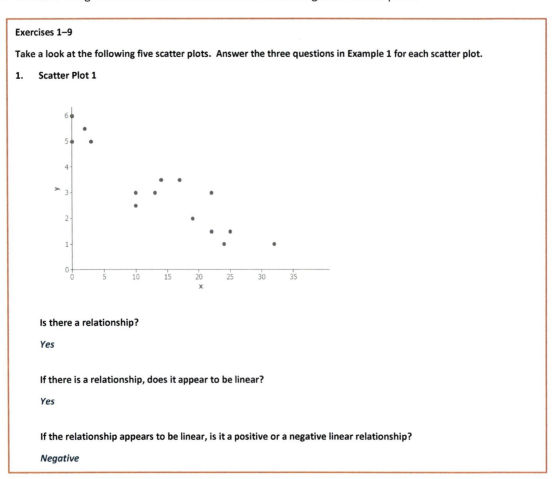

Lesson 7: Patterns in Scatter Plots

2. Scatter Plot 2

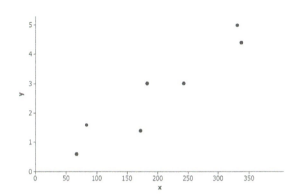

Is there a relationship?

Yes

If there is a relationship, does it appear to be linear?

Yes

If the relationship appears to be linear, is it a positive or a negative linear relationship?

Positive

3. Scatter Plot 3

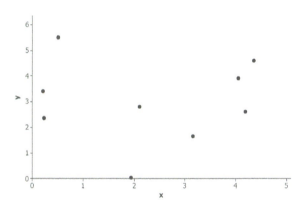

Is there a relationship?

No

If there is a relationship, does it appear to be linear?

Not applicable

If the relationship appears to be linear, is it a positive or a negative linear relationship?

Not applicable

4. **Scatter Plot 4**

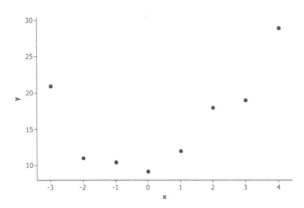

Is there a relationship?

Yes

If there is a relationship, does it appear to be linear?

No

If the relationship appears to be linear, is it a positive or a negative linear relationship?

Not applicable

5. **Scatter Plot 5**

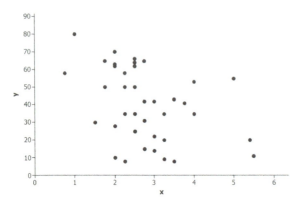

Is there a relationship?

Yes

If there is a relationship, does it appear to be linear?

Yes

If the relationship appears to be linear, is it a positive or a negative linear relationship?

Negative

A STORY OF RATIOS   Lesson 7   8•6

Exercises 6–9 (10 minutes)

Let students work in pairs on Exercises 6–9. Encourage students to use terms such as *linear* and *nonlinear* and *positive* and *negative* in their descriptions. Also, remind students that their descriptions should be written making use of the context of the problem. Point out that a good description would provide answers to the three questions they answered in the previous exercises.

> **Scaffolding:**
> - It may be helpful to provide sentence frames on the classroom board to help students articulate their observations.
> - For example, "I see a negative or positive linear relationship between ____ and ____. The higher or lower the ____, the higher or lower the ____."

6. Below is a scatter plot of data on weight in pounds (x) and fuel efficiency in miles per gallon (y) for 13 cars. Using the questions at the beginning of this lesson as a guide, write a few sentences describing any possible relationship between x and y.

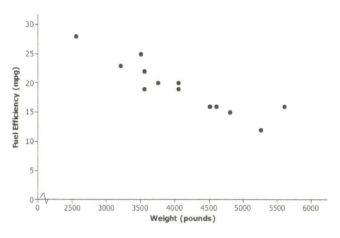

Possible response: There appears to be a negative linear relationship between fuel efficiency and weight. Students may note that this is a fairly strong negative relationship. The cars with greater weight tend to have lesser fuel efficiency.

7. Below is a scatter plot of data on price in dollars (x) and quality rating (y) for 14 bike helmets. Using the questions at the beginning of this lesson as a guide, write a few sentences describing any possible relationship between x and y.

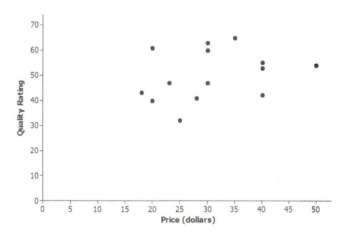

Possible response: There does not appear to be a relationship between quality rating and price. The points in the scatter plot appear to be scattered at random, and there is no apparent pattern in the scatter plot.

Lesson 7:   Patterns in Scatter Plots

8. Below is a scatter plot of data on shell length in millimeters (x) and age in years (y) for 27 lobsters of known age. Using the questions at the beginning of this lesson as a guide, write a few sentences describing any possible relationship between x and y.

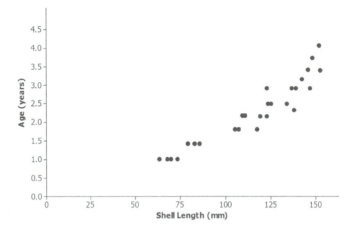

Possible response: There appears to be a relationship between shell length and age, but the pattern in the scatter plot is curved rather than linear. Age appears to increase as shell length increases, but the increase is not at a constant rate.

9. Below is a scatter plot of data from crocodiles on body mass in pounds (x) and bite force in pounds (y). Using the questions at the beginning of this lesson as a guide, write a few sentences describing any possible relationship between x and y.

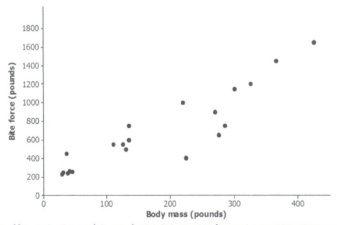

Data Source: http://journals.plos.org/plosone/article?id=10.1371/journal.pone.0031781#pone-0031781-t001

(Note: Body mass and bite force have been converted to pounds from kilograms and newtons, respectively.)

Possible response: There appears to be a positive linear relationship between bite force and body mass. For crocodiles, the greater the body mass, the greater the bite force tends to be. Students may notice that this is a positive relationship but not quite as strong as the relationship noted in Exercise 6.

Example 2 (5 minutes): Clusters and Outliers

Spend a few minutes introducing the meaning of the terms *clusters* and *outliers* in the context of scatter plots. Consider asking students to sketch a scatter plot that has an outlier and a scatter plot that has two clusters as a way of checking their understanding of these terms before moving on to the exercises that follow.

Scaffolding:
English language learners need the chance to practice using the terms *clusters* and *outliers* in both oral and written contexts. Sentence frames may be useful for students to communicate initial ideas.

> **Example 2: Clusters and Outliers**
>
> In addition to looking for a general pattern in a scatter plot, you should also look for other interesting features that might help you understand the relationship between two variables. Two things to watch for are as follows:
>
> - **CLUSTERS:** Usually, the points in a scatter plot form a single cloud of points, but sometimes the points may form two or more distinct clouds of points. These clouds are called *clusters*. Investigating these clusters may tell you something useful about the data.
> - **OUTLIERS:** An *outlier* is an unusual point in a scatter plot that does not seem to fit the general pattern or that is far away from the other points in the scatter plot.
>
> The scatter plot below was constructed using data from a study of Rocky Mountain elk ("Estimating Elk Weight from Chest Girth," *Wildlife Society Bulletin*, 1996). The variables studied were chest girth in centimeters (x) and weight in kilograms (y).
>
>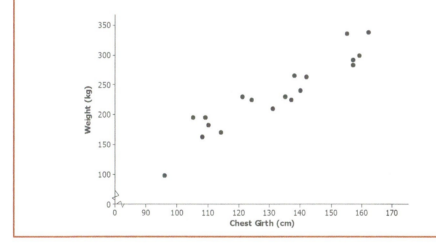

Scaffolding:
- The terms *elk* and *girth* may not be familiar to English language learners.
- An *elk* is a large mammal, similar to a deer.
- *Girth* refers to the measurement around something. For this problem, *girth* refers to the measurement around the elk from behind the front legs and under the belly. A visual aid of an elk (found on several websites) would help explain an elk's chest girth.
- Consider providing students with sentence frames or word banks, and allow students to respond in their first language to these exercises.

Exercises 10–12 (8 minutes)

Have students work individually or in pairs on Exercises 10–12. Then, have students share answers to these exercises, and discuss any of the exercises where there is disagreement on the answers.

> **Exercises 10–12**
>
> 10. Do you notice any point in the scatter plot of elk weight versus chest girth that might be described as an outlier? If so, which one?
>
>   *Possible response: The point in the lower left-hand corner of the plot corresponding to an elk with a chest girth of about 96 cm and a weight of about 100 kg could be described as an outlier. There are no other points in the scatter plot that are near this one.*

A STORY OF RATIOS　　　　　　　　　　　　　　　　　　　　　　　Lesson 7　8•6

11. If you identified an outlier in Exercise 10, write a sentence describing how this data observation differs from the others in the data set.

    Possible response: This point corresponds to an observation for an elk that is much smaller than the other elk in the data set, both in terms of chest girth and weight.

12. Do you notice any clusters in the scatter plot? If so, how would you distinguish between the clusters in terms of chest girth? Can you think of a reason these clusters might have occurred?

    Possible response: Other than the outlier, there appear to be three clusters of points. One cluster corresponds to elk with chest girths between about 105 cm and 115 cm. A second cluster includes elk with chest girths between about 120 cm and 145 cm. The third cluster includes elk with chest girths above 150 cm. It may be that age and sex play a role. Maybe the cluster with the smaller chest girths includes young elk. The two other clusters might correspond to females and males if there is a difference in size for the two sexes for Rocky Mountain elk. If we had data on age and sex, we could investigate this further.

Closing (3–5 minutes)

Consider posing the following questions; allow a few student responses for each.

- Why do you think it is a good idea to look at a scatter plot when you have data on two numerical variables?
    - *Possible response: Looking at a scatter plot makes it easier to see if there is a relationship between the two variables. It is hard to determine if there is a relationship when you just have the data in a table or a list.*
- What should you look for when you are looking at a scatter plot?
    - *Possible response: First, you should look for any general patterns. If there are patterns, you then want to consider whether the pattern is linear or nonlinear, and if it is linear, whether the relationship is positive or negative. Finally, it is also a good idea to look for any other interesting features such as outliers or clusters. The closer the points are to a line, the "stronger" the linear relationship.*

> *Scaffolding:*
> Allowing English language learners to brainstorm with a partner first may elicit a greater response in the whole-group setting.

Lesson Summary

- A scatter plot might show a linear relationship, a nonlinear relationship, or no relationship.
- A positive linear relationship is one that would be modeled using a line with a positive slope. A negative linear relationship is one that would be modeled by a line with a negative slope.
- Outliers in a scatter plot are unusual points that do not seem to fit the general pattern in the plot or that are far away from the other points in the scatter plot.
- Clusters occur when the points in the scatter plot appear to form two or more distinct clouds of points.

Exit Ticket (5 minutes)

Lesson 7:   Patterns in Scatter Plots

Lesson 7: Patterns in Scatter Plots

Exit Ticket

1. Which of the following scatter plots shows a negative linear relationship? Explain how you know.

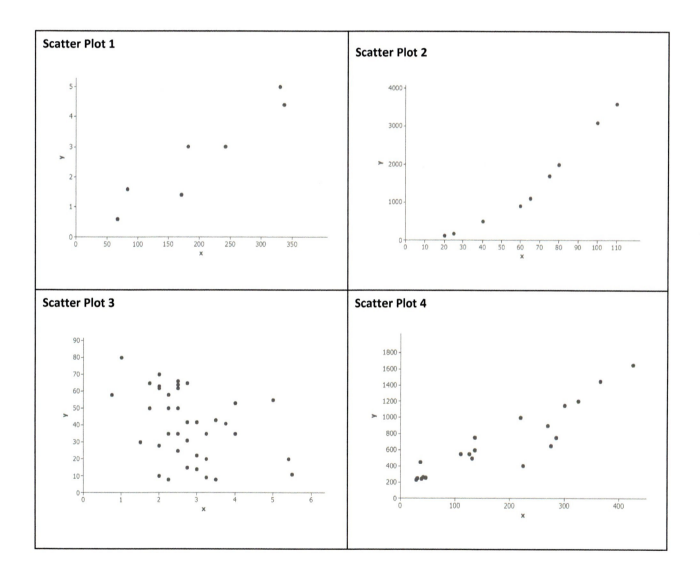

2. The scatter plot below was constructed using data from eighth-grade students on number of hours playing video games per week (x) and number of hours of sleep per night (y). Write a few sentences describing the relationship between sleep time and time spent playing video games for these students. Are there any noticeable clusters or outliers?

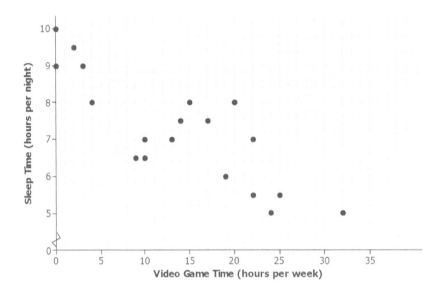

3. In a scatter plot, if the values of y tend to increase as the value of x increases, would you say that there is a positive relationship or a negative relationship between x and y? Explain your answer.

Exit Ticket Sample Solutions

1. Which of the following scatter plots shows a negative linear relationship? Explain how you know.

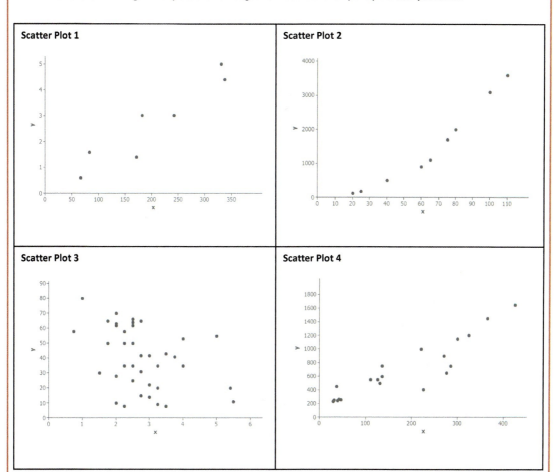

Only Scatter Plot 3 shows a negative linear relationship because the y-values tend to decrease as the value of x increases.

2. The scatter plot below was constructed using data from eighth-grade students on number of hours playing video games per week (x) and number of hours of sleep per night (y). Write a few sentences describing the relationship between sleep time and time spent playing video games for these students. Are there any noticeable clusters or outliers?

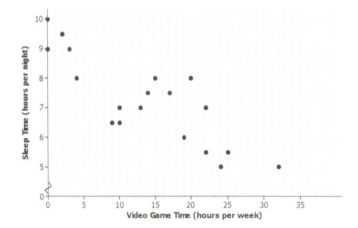

Answers will vary. Sample response: There appears to be a negative linear relationship between the number of hours per week a student plays video games and the number of hours per night the student sleeps. As video game time increases, the number of hours of sleep tends to decrease. There is one observation that might be considered an outlier—the point corresponding to a student who plays video games 32 hours per week. Other than the outlier, there are two clusters—one corresponding to students who spend very little time playing video games and a second corresponding to students who play video games between about 10 and 25 hours per week.

3. In a scatter plot, if the value of y tends to increase as the value of x increases, would you say that there is a positive relationship or a negative relationship between x and y? Explain your answer.

There is a positive relationship. If the value of y increases as the value of x increases, the points go up on the scatter plot from left to right.

Problem Set Sample Solutions

The Problem Set is intended to reinforce material from the lesson and have students think about the meaning of points in a scatter plot, clusters, positive and negative linear trends, and trends that are not linear.

1. Suppose data was collected on size in square feet (x) of several houses and price in dollars (y). The data was then used to construct the scatterplot below. Write a few sentences describing the relationship between price and size for these houses. Are there any noticeable clusters or outliers?

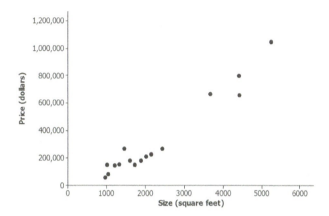

Answers will vary. Possible response: There appears to be a positive linear relationship between size and price. Price tends to increase as size increases. There appear to be two clusters of houses—one that includes houses that are less than 3,000 square feet in size and another that includes houses that are more than 3,000 square feet in size.

2. The scatter plot below was constructed using data on length in inches (x) of several alligators and weight in pounds (y). Write a few sentences describing the relationship between weight and length for these alligators. Are there any noticeable clusters or outliers?

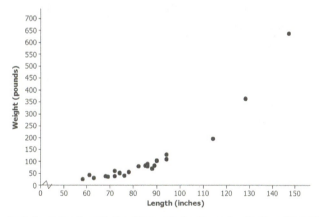

Data Source: Exploring Data, Quantitative Literacy Series, James Landwehr and Ann Watkins, 1987.

Answers will vary. Possible response: There appears to be a positive relationship between length and weight, but the relationship is not linear. Weight tends to increase as length increases. There are three observations that stand out as outliers. These correspond to alligators that are much bigger in terms of both length and weight than the other alligators in the sample. Without these three alligators, the relationship between length and weight would look linear. It might be possible to use a line to model the relationship between weight and length for alligators that have lengths of fewer than 100 inches.

Lesson 7: Patterns in Scatter Plots

3. Suppose the scatter plot below was constructed using data on age in years (x) of several Honda Civics and price in dollars (y). Write a few sentences describing the relationship between price and age for these cars. Are there any noticeable clusters or outliers?

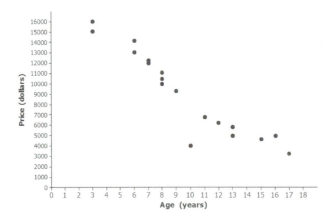

Answers will vary. Possible response: There appears to be a negative linear relationship between price and age. Price tends to decrease as age increases. There is one car that looks like an outlier—the car that is 10 years old. This car has a price that is lower than expected based on the pattern of the other points in the scatter plot.

4. Samples of students in each of the U.S. states periodically take part in a large-scale assessment called the National Assessment of Educational Progress (NAEP). The table below shows the percent of students in the northeastern states (as defined by the U.S. Census Bureau) who answered Problems 7 and 15 correctly on the 2011 eighth-grade test. The scatter plot shows the percent of eighth-grade students who got Problems 7 and 15 correct on the 2011 NAEP.

State	Percent Correct Problem 7	Percent Correct Problem 15
Connecticut	29	51
New York	28	47
Rhode Island	29	52
Maine	27	50
Pennsylvania	29	48
Vermont	32	58
New Jersey	35	54
New Hampshire	29	52
Massachusetts	35	56

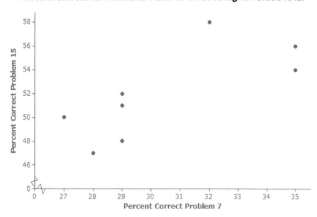

a. Why does it appear that there are only eight points in the scatter plot for nine states?

Two of the states, New Hampshire and Rhode Island, had exactly the same percent correct on each of the questions, (29, 52).

b. What is true of the states represented by the cluster of five points in the lower left corner of the graph?

Answers will vary; those states had lower percentages correct than the other three states in the upper right.

c. Which state did the best on these two problems? Explain your reasoning.

Answers will vary; some students might argue that Massachusetts at (35, 56) did the best. Even though Vermont actually did a bit better on Problem 15, it was lower on Problem 7.

d. Is there a trend in the data? Explain your thinking.

Answers will vary; there seems to be a positive linear trend, as a large percent correct on one question suggests a large percent correct on the other, and a low percent on one suggests a low percent on the other.

5. The plot below shows the mean percent of sunshine during the year and the mean amount of precipitation in inches per year for the states in the United States.

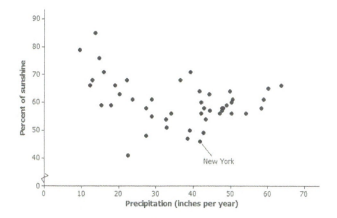

Data source: www.currentresults.com/Weather/US/average-annual-state-sunshine.php
www.currentresults.com/Weather/US/average-annual-state-precipitation.php

a. Where on the graph are the states that have a large amount of precipitation and a small percent of sunshine?

Those states will be in the lower right-hand corner of the graph.

b. The state of New York is the point $(46, 41.8)$. Describe how the mean amount of precipitation and percent of sunshine in New York compare to the rest of the United States.

New York has a little over 40 inches of precipitation per year and is sunny about 45% of the time. It has a smaller percent of sunshine over the year than most states and is about in the middle of the states in terms of the amount of precipitation, which goes from about 10 to 65 inches per year.

c. Write a few sentences describing the relationship between mean amount of precipitation and percent of sunshine.

There is a negative relationship, or the more precipitation, the less percent of sun. If you took away the three states at the top left with a large percent of sun and very little precipitation, the trend would not be as pronounced. The relationship is not linear.

6. At a dinner party, every person shakes hands with every other person present.

   a. If three people are in a room and everyone shakes hands with everyone else, how many handshakes take place?

   Three handshakes

   b. Make a table for the number of handshakes in the room for one to six people. You may want to make a diagram or list to help you count the number of handshakes.

Number People	Handshakes
1	0
2	1
3	3

Number People	Handshakes
4	6
5	10
6	15

   c. Make a scatter plot of number of people (x) and number of handshakes (y). Explain your thinking.

   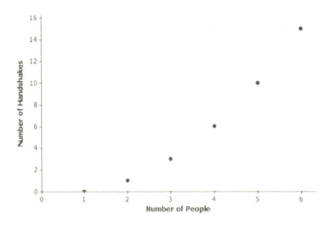

   d. Does the trend seem to be linear? Why or why not?

   The trend is increasing, but it is not linear. As the number of people increases, the number of handshakes also increases. It does not increase at a constant rate.

A STORY OF RATIOS   Lesson 8   8•6

 Lesson 8: Informally Fitting a Line

Student Outcomes

- Students informally fit a straight line to data displayed in a scatter plot.
- Students make predictions based on the graph of a line that has been fit to data.

Lesson Notes

In this lesson, students investigate scatter plots of data and informally fit a line to the pattern observed in the plot. Students then make predictions based on their lines. Students informally evaluate their predictions based on the fit of the line to the data.

Classwork

Example 1 (2–3 minutes): Housing Costs

Introduce the data presented in the table and the scatter plot of the data. Ask students the following:

- Examine the scatter plot. What trend do you see? How would you describe this trend?
  - *It appears to be a positive linear trend. The scatter plot indicates that the larger the size, the higher the price.*

Scaffolding:
- The terms *house* and *home* are used interchangeably throughout the example.
- This may be confusing for English language learners and should be clarified.

(Note: Make sure to give students an opportunity to explain why they think there is a positive linear trend between price and size.)

Example 1: Housing Costs

Let's look at some data from one midwestern city that indicate the sizes and sale prices of various houses sold in this city.

Size (square feet)	Price (dollars)	Size (square feet)	Price (dollars)
5,232	1,050,000	1,196	144,900
1,875	179,900	1,719	149,900
1,031	84,900	956	59,900
1,437	269,900	991	149,900
4,400	799,900	1,312	154,900
2,000	209,900	4,417	659,999
2,132	224,900	3,664	669,000
1,591	179,900	2,421	269,900

Data Source: http://www.trulia.com/for_sale/Milwaukee,WI/5_p, accessed in 2013

   Lesson 8:   Informally Fitting a Line

A STORY OF RATIOS   Lesson 8   8•6

A scatter plot of the data is given below.

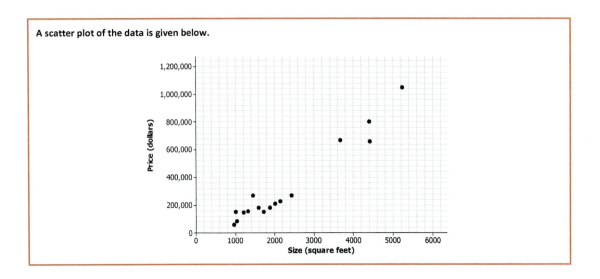

Exercises 1–6 (15 minutes)

In these exercises, be sure that students retain the units as they write and discuss the solutions, being mindful of attending to precision. Students might use a transparent ruler or a piece of uncooked spaghetti to help draw and decide where to place their lines. To avoid problems with the size of the numbers and to have students focus on drawing their lines, the teacher should provide a worksheet for students with the points already plotted on a grid. Students should concentrate on the general form of the scatter plot rather than worrying too much about the exact placement of points in the scatter plot. The primary focus of the work in these exercises is to have students think about the trend, use a line to describe the trend, and make predictions based on the line.

Work through the exercises as a class, allowing time to discuss multiple responses.

Exercises 1–6

1. What can you tell about the price of large homes compared to the price of small homes from the table?

   Answers will vary. Students should make the observation that, overall, the larger homes cost more, and the smaller homes cost less. However, it is hard to generalize because one of the smaller homes costs nearly $150,000.

2. Use the scatter plot to answer the following questions.

   a. Does the scatter plot seem to support the statement that larger houses tend to cost more? Explain your thinking.

   Yes, because the trend is positive, the larger the size of the house, the more the house tends to cost.

   b. What is the cost of the most expensive house, and where is that point on the scatter plot?

   The house with a size of 5,232 square feet costs $1,050,000, which is the most expensive. It is in the upper right corner of the scatter plot.

96   Lesson 8:   Informally Fitting a Line

A STORY OF RATIOS   Lesson 8   8•6

c. Some people might consider a given amount of money and then predict what size house they could buy. Others might consider what size house they want and then predict how much it would cost. How would you use the scatter plot in Example 1?

Answers will vary. Since the size of the house is on the horizontal axis and the price is on the vertical axis, the scatter plot is set up with price as the dependent variable and size as the independent variable. This is the way you would set it up if you wanted to predict price based on size. Although various answers are appropriate, move the discussion along using size to predict price.

d. Estimate the cost of a 3,000-square-foot house.

Answers will vary. Reasonable answers range between $300,000 and $600,000.

e. Do you think a line would provide a reasonable way to describe how price and size are related? How could you use a line to predict the price of a house if you are given its size?

Answers will vary; however, use this question to develop the idea that a line would provide a way to estimate the cost given the size of a house. The challenge is how to make that line. Note: Students are encouraged in the next exercise to first make a line and then evaluate whether or not it fits the data. This will provide a reasonable estimate of the cost of a house in relation to its size.

3. Draw a line in the plot that you think would fit the trend in the data.

Answers will vary. Discuss several of the lines students have drawn by encouraging students to share their lines with the class. At this point, do not evaluate the lines as good or bad. Students may want to know a precise procedure or process to draw their lines. If that question comes up, indicate to students that a procedure will be developed in their future work (Algebra I) with statistics. For now, the goal is to simply draw a line that can be used to describe the relationship between the size of a home and its cost. Indicate that strategies for drawing a line will be explored in Exercise 5. Use the lines provided by students to evaluate the predictions in the following exercise. These predictions are used to develop a strategy for drawing a line. Use the line drawn by students to highlight their understanding of the data.

4. Use your line to answer the following questions:
   a. What is your prediction of the price of a 3,000-square-foot house?

   Answers will vary. A reasonable prediction is around $500,000.

   b. What is the prediction of the price of a 1,500-square-foot house?

   Answers will vary. A reasonable prediction is around $200,000.

> Scaffolding:
> - Point out to students that the word *trend* is not connected to the use of this word in describing fashion or music. (For example, "the trend in music is for more use of drums.")
> - In this lesson, *trend* describes the pattern or lack of a pattern in the scatter plot.
> - Ask students to highlight words that they think would describe a trend in the scatter plots that are examined in this lesson.
> - Explain to English language learners that *scatter plot* may be referred to as just *plot*.

Display various predictions students found for these two examples. Consider using a chart similar to the following to discuss the different predictions.

Student	Estimate of the Price for a 3,000-Square-Foot House	Estimate of the Price for a 1,500-Square-Foot House
Student 1	$300,000	$100,000
Student 2	$600,000	$400,000

Lesson 8:   Informally Fitting a Line   97

©2018 Great Minds®. eureka-math.org

A STORY OF RATIOS   Lesson 8   8•6

Discuss that predictions vary as a result of the different lines that students used to describe the pattern in the scatter plot. What line makes the most sense for these data?

Before discussing answers to that question, encourage students to explain how they drew their lines and why their predictions might have been higher (or lower) than other students'. For example, students with lines that are visibly above most of the points may have predictions that are higher than the predictions of students with lines below several of the points. Ask students to summarize their theories of how to draw a line as a *strategy* for drawing a line. After they provide their own descriptions, provide students an opportunity to think about the following strategies that might have been used to draw a line.

> 5. Consider the following general strategies students use for drawing a line. Do you think they represent a good strategy for drawing a line that fits the data? Explain why or why not, or draw a line for the scatter plot using the strategy that would indicate why it is or why it is not a good strategy.
>
>   a. Laure thought she might draw her line using the very first point (farthest to the left) and the very last point (farthest to the right) in the scatter plot.
>
>   *Answers will vary. This may work in some cases, but those points might not capture the trend in the data. For example, the first point in the lower left might not be in line with the other points.*
>
>   b. Phil wants to be sure that he has the same number of points above and below the line.
>
>   *Answers will vary. You could draw a nearly horizontal line that has half of the points above and half below, but that might not represent the trend in the data at all. Note: For many students just starting out, this seems like a reasonable strategy, but it often can result in lines that clearly do not fit the data. As indicated, drawing a nearly horizontal line is a good way to indicate that this is not a good strategy.*
>
>   c. Sandie thought she might try to get a line that had the most points right on it.
>
>   *Answers will vary. That might result in, perhaps, three points on the line (knowing it only takes two to make a line), but the others could be anywhere. The line might even go in the wrong direction. Note: For students just beginning to think of how to draw a line, this seems like a reasonable goal; however, point out that this strategy may result in lines that are not good for predicting price.*
>
>   d. Maree decided to get her line as close to as many of the points as possible.
>
>   *Answers will vary. If you can figure out how to do this, Maree's approach seems like a reasonable way to find a line that takes all of the points into account.*
>
> 6. Based on the strategies discussed in Exercise 5, would you change how you draw a line through the points? Explain your answer.
>
>   *Answers will vary based on how a student drew his original line. Summarize that the goal is to draw a line that is as close as possible to the points in the scatter plot. More precise methods are developed in Algebra I.*

Example 2 (2–3 minutes): Deep Water

Introduce students to the data in the table. Pose the questions in the text, and allow for multiple responses.

> **Example 2: Deep Water**
>
> Does the current in the water go faster or slower when the water is shallow? The data on the depth and velocity of the Columbia River at various locations in Washington State listed on the next page can help you think about the answer.

98   Lesson 8:   Informally Fitting a Line

Depth and Velocity in the Columbia River, Washington State

Depth (feet)	Velocity (feet/second)
0.7	1.55
2.0	1.11
2.6	1.42
3.3	1.39
4.6	1.39
5.9	1.14
7.3	0.91
8.6	0.59
9.9	0.59
10.6	0.41
11.2	0.22

Data Source: www.seattlecentral.edu/qelp/sets/011/011.html

Scaffolding:
- The word *current* has multiple meanings that English language learners may be familiar with from a social studies class (current events) or from a science class (electrical current).
- In this example, *current* refers to the flow or velocity of the river.

a. What can you tell about the relationship between the depth and velocity by looking at the numbers in the table?

Answers will vary. According to the table, as the depth increases, the velocity appears to decrease.

b. If you were to make a scatter plot of the data, which variable would you put on the horizontal axis, and why?

Answers will vary. It might be easier to measure the depth and use that information to predict the velocity of the water, so the depth should go on the horizontal axis.

Exercises 7–9 (12–15 minutes)

These exercises engage students in a context where the trend has a negative slope. Again, students should pay careful attention to units and interpretation of rate of change. They evaluate the line by assessing its closeness to the data points. Let students work with a partner. If time allows, discuss the answers as a class.

Exercises 7–9

7. A scatter plot of the Columbia River data is shown below.

Scaffolding:
English language learners may need support in recognizing the relationship between the words *depth* and *deep*.

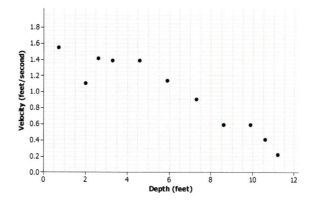

a. Choose a data point in the scatter plot, and describe what it means in terms of the context.

Answers will vary. For example, $(4.6, 1.39)$ would represent a place in the river that was 4.6 feet deep and had a velocity of 1.39 ft/sec.

Lesson 8: Informally Fitting a Line

b. Based on the scatter plot, describe the relationship between velocity and depth.

The deeper the water, the slower the current velocity tends to be.

c. How would you explain the relationship between the velocity and depth of the water?

Answers will vary. Sample response: Velocity may be a result of the volume of water. Shallow water has less volume, and as a result, the water runs faster. Note: Students may have several explanations. For example, they may say that depth is a result of less water runoff; therefore, water depth increases.

d. If the river is two feet deep at a certain spot, how fast do you think the current would be? Explain your reasoning.

Answers will vary. Based on the data, it could be around 1.11 ft/sec, or it could be closer to 1.42 ft/sec, which is more in line with the pattern for the other points.

8. Consider the following questions:

   a. If you draw a line to represent the trend in the plot, would it make it easier to predict the velocity of the water if you know the depth? Why or why not?

   Answers will vary. A line will help you determine a better prediction for 1.5 ft. or 5 ft., where the points are a bit scattered.

   b. Draw a line that you think does a reasonable job of modeling the trend on the scatter plot in Exercise 7. Use the line to predict the velocity when the water is 8 feet deep.

   Answers will vary. A line is drawn in the following graph. Using this line, when the water is 8 ft. deep, the velocity is predicted to be 0.76 ft/sec.

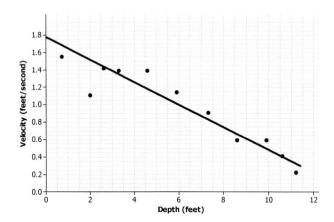

9. Use the line to predict the velocity for a depth of 8.6 feet. How far off was your prediction from the actual observed velocity for the location that had a depth of 8.6 feet?

Answers will vary. Sample response: The current would be moving at a velocity of 0.68 ft/sec. The observed velocity was 0.59 ft/sec, so the line predicted a velocity that was 0.09 ft/sec faster than the observed value.

Lesson 8: Informally Fitting a Line

A STORY OF RATIOS   Lesson 8   8•6

Closing (5 minutes)

Consider posing the following questions; allow a few student responses for each.

- How do scatter plots and tables of data differ in helping you understand the "story" when looking at bivariate numerical data?
    - *The numbers in a table can give you a sense of how big or small the values are, but it is easier to see a relationship between the variables in a scatter plot.*
- What is the difference between predicting an outcome by looking at a scatter plot and predicting the outcome using a line that models the trend?
    - *When you look at the plot, the points are sometimes very spread out, and for a given value of an independent variable, some values you might be interested in may not be included in the data set. Using a line takes all of the points into consideration, and your prediction is based on an overall pattern rather than just one or two points.*
- In a scatter plot, which variable goes on the horizontal axis, and which goes on the vertical axis?
    - *The independent variable (or the variable not changed by other variables) goes on the horizontal axis, and the dependent variable (or the variable to be predicted by the independent variable) goes on the vertical axis.*

Lesson Summary

- When constructing a scatter plot, the variable that you want to predict (i.e., the dependent or response variable) goes on the vertical axis. The independent variable (i.e., the variable not related to other variables) goes on the horizontal axis.
- When the pattern in a scatter plot is approximately linear, a line can be used to describe the linear relationship.
- A line that describes the relationship between a dependent variable and an independent variable can be used to make predictions of the value of the dependent variable given a value of the independent variable.
- When informally fitting a line, you want to find a line for which the points in the scatter plot tend to be closest.

Exit Ticket (5 minutes)

Lesson 8:   Informally Fitting a Line

Name _____   Date_____

Lesson 8: Informally Fitting a Line

Exit Ticket

The plot below is a scatter plot of mean temperature in July and mean inches of rain per year for a sample of midwestern cities. A line is drawn to fit the data.

July Temperatures and Rainfall in Selected Midwestern Cities

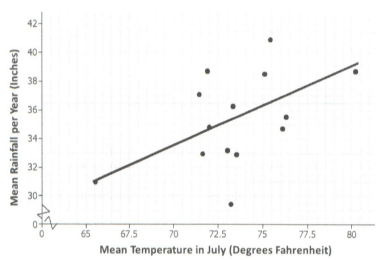

Data Source: http://countrystudies.us/united-states/weather/

1. Choose a point in the scatter plot, and explain what it represents.

2. Use the line provided to predict the mean number of inches of rain per year for a city that has a mean temperature of 70°F in July.

3. Do you think the line provided is a good one for this scatter plot? Explain your answer.

Exit Ticket Sample Solutions

The plot below is a scatter plot of mean temperature in July and mean inches of rain per year for a sample of midwestern cities. A line is drawn to fit the data.

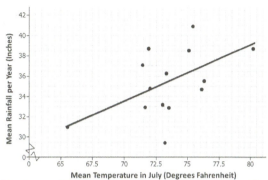

Data Source: http://countrystudies.us/united-states/weather/

1. Choose a point in the scatter plot, and explain what it represents.

   Answers will vary. Sample response: The point at about (72, 35) represents a Midwestern city where the mean temperature in July is about 72°F and where the rainfall per year is about 35 inches.

2. Use the line provided to predict the mean number of inches of rain per year for a city that has a mean temperature of 70°F in July.

   Predicted rainfall is 33 inches of rain per year. (Some students will state approximately 33.5 inches of rain.)

3. Do you think the line provided is a good one for this scatter plot? Explain your answer.

   Yes. The line follows the general pattern in the scatter plot, and it does not look like there is another area in the scatter plot where the points would be any closer to the line.

Problem Set Sample Solutions

1. The table below shows the mean temperature in July and the mean amount of rainfall per year for 14 cities in the Midwest.

City	Mean Temperature in July (degrees Fahrenheit)	Mean Rainfall per Year (inches)
Chicago, IL	73.3	36.27
Cleveland, OH	71.9	38.71
Columbus, OH	75.1	38.52
Des Moines, IA	76.1	34.72
Detroit, MI	73.5	32.89
Duluth, MN	65.5	31.00
Grand Rapids, MI	71.4	37.13
Indianapolis, IN	75.4	40.95
Marquette, MI	71.6	32.95
Milwaukee, WI	72.0	34.81
Minneapolis–St. Paul, MN	73.2	29.41
Springfield, MO	76.3	35.56
St. Louis, MO	80.2	38.75
Rapid City, SD	73.0	33.21

Data Source: http://countrystudies.us/united-states/weather/

a. What do you observe from looking at the data in the table?

Answers will vary. Many of the temperatures were in the 70's, and many of the mean inches of rain were in the 30's. It also appears that, in general, as the rainfall increased, the mean temperature also increased.

b. Look at the scatter plot below. A line is drawn to fit the data. The plot in the Exit Ticket had the mean July temperatures for the cities on the horizontal axis. How is this plot different, and what does it mean for the way you think about the relationship between the two variables—temperature and rain?

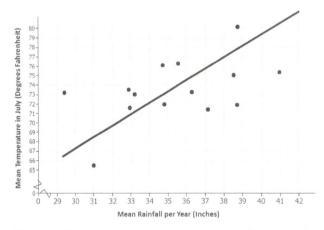

This scatter plot has the labels on the axes reversed: (mean inches of rain, mean temperature). This is the scatter plot I would use if I wanted to predict the mean temperature in July knowing the mean amount of rain per year.

c. The line has been drawn to model the relationship between the amount of rain and the temperature in those Midwestern cities. Use the line to predict the mean July temperature for a Midwestern city that has a mean of 32 inches of rain per year.

Answers will vary. For 32 in. of rain per year, the line indicates a mean July temperature of approximately 70°F.

d. For which of the cities in the sample does the line do the worst job of predicting the mean temperature? The best? Explain your reasoning with as much detail as possible.

Answers will vary. I looked for points that were really close to the line and ones that were far away. The line prediction for temperature would be farthest off for Minneapolis. For 29.41 in. of rain in Minneapolis, the line predicted approximately 67°F, whereas the actual mean temperature in July was 73.2°F. The line predicted very well for Milwaukee. For 32.95 in. of rain in Milwaukee, the line predicted approximately 73°F, whereas the actual mean temperature in July was 72°F and was only off by about 1°F. The line was also close for Marquette. For 34.81 in. of rain in Marquette, the line predicted approximately 71°F, whereas the actual mean temperature in July was 71.6°F and was only off by about 1°F.

2. The scatter plot below shows the results of a survey of eighth-grade students who were asked to report the number of hours per week they spend playing video games and the typical number of hours they sleep each night.

Mean Hours Sleep per Night Versus Mean Hours Playing Video Games per Week

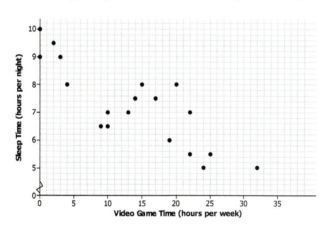

a. What trend do you observe in the data?

The more hours that students play video games, the fewer hours they tend to sleep.

b. What was the fewest number of hours per week that students who were surveyed spent playing video games? The most?

Two students spent 0 hours, and one student spent 32 hours per week playing games.

c. What was the fewest number of hours per night that students who were surveyed typically slept? The most?

The fewest hours of sleep per night was around 5 hours, and the most was around 10 hours.

d. Draw a line that seems to fit the trend in the data, and find its equation. Use the line to predict the number of hours of sleep for a student who spends about 15 hours per week playing video games.

Answers will vary. A student who spent 15 hours per week playing games would get about 7 hours of sleep per night.

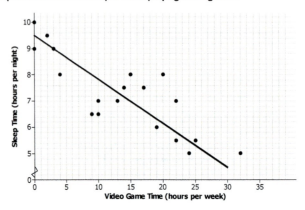

3. Scientists can take very good pictures of alligators from airplanes or helicopters. Scientists in Florida are interested in studying the relationship between the length and the weight of alligators in the waters around Florida.

   a. Would it be easier to collect data on length or weight? Explain your thinking.

   Answers will vary. You could measure the length from the pictures, but you would have to actually have the alligators to weigh them.

   b. Use your answer to decide which variable you would want to put on the horizontal axis and which variable you might want to predict.

   You would probably want to predict the weight of the alligator knowing the length; therefore, the length would go on the horizontal axis and the weight on the vertical axis.

4. Scientists captured a small sample of alligators and measured both their length (in inches) and weight (in pounds). Torre used their data to create the following scatter plot and drew a line to capture the trend in the data. She and Steve then had a discussion about the way the line fit the data. What do you think they were discussing, and why?

Alligator Length (inches) and Weight (pounds)

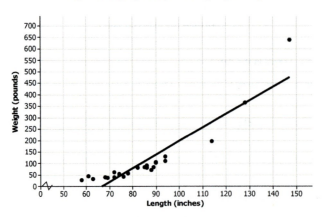

Data Source: James Landwehr and Ann Watkins, *Exploring Data*, Quantitative Literacy Series (Dale Seymour, 1987).

Answers will vary. Sample response: The pattern in the scatter plot is curved instead of linear. All of the data points in the middle of the scatter plot fall below the line, and the line does not really capture the pattern in the scatter plot. A line does not pass through the cluster of points between 60 to 80 in. in length that fit the other points. A model other than a line might be a better fit.

Lesson 9: Determining the Equation of a Line Fit to Data

Student Outcomes

- Students informally fit a straight line to data displayed in a scatter plot.
- Students determine the equation of a line fit to data.
- Students make predictions based on the equation of a line fit to data.

Lesson Notes

In this lesson, students informally fit a line to data by drawing a line that describes a linear pattern in a scatter plot and then use their lines to make predictions. They determine the equation of the line and informally analyze different lines fit to the same data. This lesson begins developing the foundation for finding an objective way to judge how well a line fits the trend in a scatter plot and the notion of a *best-fit* line in Algebra I.

Classwork

Example 1 (5 minutes): Crocodiles and Alligators

Discuss the data presented in the table and scatter plot. Consider starting by asking if students are familiar with crocodiles and alligators and how they differ. Ask students if they can imagine what a bite force of 100 pounds would feel like. Ask them if they know what body mass indicates. If students understand that body mass is an indication of the weight of a crocodilian and bite force is a measure of the strength of a crocodilian's bite, the data can be investigated even if they do not understand the technical definitions and how these variables are measured. Also, ask students if any other aspects of the data surprised them. For example, did they realize that there are so many different species of crocodilians? Did the wide range of body mass and bite force surprise them? If time permits, suggest that students do further research on crocodilians.

Example 1: Crocodiles and Alligators

Scientists are interested in finding out how different species adapt to finding food sources. One group studied crocodilians to find out how their bite force was related to body mass and diet. The table below displays the information they collected on body mass (in pounds) and bite force (in pounds).

Crocodilian Biting

Species	Body Mass (pounds)	Bite Force (pounds)
Dwarf crocodile	35	450
Crocodile F	40	260
Alligator A	30	250
Caiman A	28	230
Caiman B	37	240
Caiman C	45	255
Crocodile A	110	550
Nile crocodile	275	650
Crocodile B	130	500
Crocodile C	135	600
Crocodile D	135	750
Caiman D	125	550
Indian gharial crocodile	225	400
Crocodile G	220	1,000
American crocodile	270	900
Crocodile E	285	750
Crocodile F	425	1,650
American alligator	300	1,150
Alligator B	325	1,200
Alligator C	365	1,450

Scaffolding:
- The word *crocodilian* refers to any reptile of the order Crocodylia.
- This includes crocodiles, alligators, caimans, and gavials. Showing students a visual aid with pictures of these animals may help them understand.

Data Source: http://journals.plos.org/plosone/article?id=10.1371/journal.pone.0031781#pone-0031781-t001

(Note: Body mass and bite force have been converted to pounds from kilograms and newtons, respectively.)

As you learned in the previous lesson, it is a good idea to begin by looking at what a scatter plot tells you about the data. The scatter plot below displays the data on body mass and bite force for the crocodilians in the study.

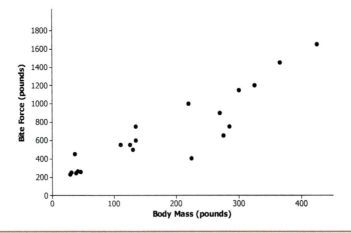

A STORY OF RATIOS   Lesson 9   8•6

Exercises 1–5 (14 minutes)

Exercises 1 through 5 ask students to consider the fit of a line. Each student (or small group of students) draws a line that would be a good representation of the trend in the data. Students evaluate their lines and the lines of the four students introduced in Exercise 4.

In Exercise 2, students draw a line they think is a good representation of the trend in the data. Ask them to compare their lines with other students'. As a group, decide who might have the best line, and ask students why they made that choice. Have groups share their ideas. Point out that it would be helpful to agree on a standard method for judging the fit of a line. One method is to look at how well the line predicts for the given data or how often it is over or under the actual or observed value.

Scaffolding:
Point out to English language learners that the terms body mass *and* weight *are used interchangeably in this lesson.*

Exercises 1–6

1. Describe the relationship between body mass and bite force for the crocodilians shown in the scatter plot.

   As the body mass increases, the bite force tends to also increase.

2. Draw a line to represent the trend in the data. Comment on what you considered in drawing your line.

   The line should be as close as possible to the points in the scatter plot. Students explored this idea in Lesson 8.

3. Based on your line, predict the bite force for a crocodilian that weighs 220 pounds. How does this prediction compare to the actual bite force of the 220-pound crocodilian in the data set?

   Answers will vary. A reasonable prediction is around 650 to 700 pounds. The actual bite force was 1,000 pounds, so the prediction based on the line was not very close for this crocodilian.

4. Several students decided to draw lines to represent the trend in the data. Consider the lines drawn by Sol, Patti, Marrisa, and Taylor, which are shown below.

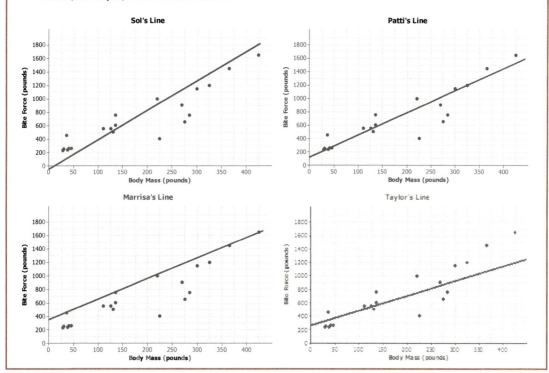

Lesson 9:   Determining the Equation of a Line Fit to Data   109

For each student, indicate whether or not you think the line would be a good line to use to make predictions. Explain your thinking.

   a. Sol's line

   In general, it looks like Sol's line overestimates the bite force for heavier crocodilians and underestimates the bite force for crocodilians that do not weigh as much.

   b. Patti's line

   Patti's line looks like it fits the data well, so it would probably produce good predictions. The line goes through the middle of the points in the scatter plot, and the points are fairly close to the line.

   c. Marrisa's line

   It looks like Marrisa's line overestimates the bite force because almost all of the points are below the line.

   d. Taylor's line

   It looks like Taylor's line tends to underestimate the bite force. There are many points above the line.

5. What is the equation of your line? Show the steps you used to determine your line. Based on your equation, what is your prediction for the bite force of a crocodilian weighing 200 pounds?

   Answers will vary. Students have learned from previous modules how to find the equation of a line. Anticipate students to first determine the slope based on two points on their lines. Students then use a point on the line to obtain an equation in the form $y = mx + b$ (or $y = a + bx$). Students use their lines to predict a bite force for a crocodilian that weighs 200 pounds. A reasonable answer would be around 800 pounds.

6. Patti drew vertical line segments from two points to the line in her scatter plot. The first point she selected was for a dwarf crocodile. The second point she selected was for an Indian gharial crocodile.

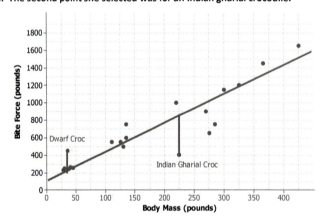

   a. Would Patti's line have resulted in a predicted bite force that was closer to the actual bite force for the dwarf crocodile or for the Indian gharial crocodile? What aspect of the scatter plot supports your answer?

   The prediction would be closer to the actual bite force for the dwarf crocodile. That point is closer to the line (the vertical line segment connecting it to the line is shorter) than the point for the Indian gharial crocodile.

   b. Would it be preferable to describe the trend in a scatter plot using a line that makes the differences in the actual and predicted values large or small? Explain your answer.

   It would be better for the differences to be as small as possible. Small differences are closer to the line.

A STORY OF RATIOS — Lesson 9 — 8•6

Exercise 7 (14 minutes): Used Cars

This exercise provides additional practice for students. Students use the equation of a line to make predictions and informally assess the fit of the line.

Exercise 7: Used Cars

7. Suppose the plot below shows the age (in years) and price (in dollars) of used compact cars that were advertised in a local newspaper.

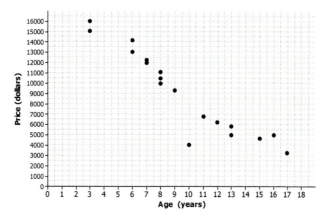

a. Based on the scatter plot above, describe the relationship between the age and price of the used cars.

The older the car, the lower the price tends to be.

b. Nora drew a line she thought was close to many of the points and found the equation of the line. She used the points $(13, 6000)$ and $(7, 12000)$ on her line to find the equation. Explain why those points made finding the equation easy.

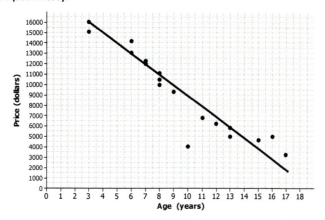

The points are at the intersection of the grid lines in the graph, so it is easy to determine the coordinates of these points.

c. Find the equation of Nora's line for predicting the price of a used car given its age. Summarize the trend described by this equation.

Using the points, the equation is $y = -1000x + 19000$, or Price $= -1000(\text{age}) + 19000$. The slope of the line is negative, so the line indicates that the price of used cars decreases as cars get older.

Lesson 9: Determining the Equation of a Line Fit to Data

d. Based on the line, for which car in the data set would the predicted value be farthest from the actual value? How can you tell?

It would be farthest for the car that is 10 years old. It is the point in the scatter plot that is farthest from the line.

e. What does the equation predict for the cost of a 10-year-old car? How close was the prediction using the line to the actual cost of the 10-year-old car in the data set? Given the context of the data set, do you think the difference between the predicted price and the actual price is large or small?

The line predicts a 10-year-old car would cost about $9,000. $-1000(10) + 19000 = 9000$. Compared to $4,040 for the 10-year-old car in the data set, the difference would be $4,960. The prediction is off by about $5,000, which seems like a lot of money, given the prices of the cars in the data set.

f. Is $5,000 typical of the differences between predicted prices and actual prices for the cars in this data set? Justify your answer.

No, most of the differences would be much smaller than $5,000. Most of the points are much closer to the line, and most predictions would be within about $1,000 of the actual value.

Closing (2 minutes)

- When you use a line to describe a linear relationship in a data set, what are characteristics of a good fit?
  - *The line should be as close as possible to the points in the scatter plot. The line should go through the "middle" of the points.*

Lesson Summary

- A line can be used to represent the trend in a scatter plot.
- Evaluating the equation of the line for a value of the independent variable determines a value predicted by the line.
- A good line for prediction is one that goes through the middle of the points in a scatter plot and for which the points tend to fall close to the line.

Exit Ticket (5 minutes)

Name _____ Date _____

Lesson 9: Determining the Equation of a Line Fit to Data

Exit Ticket

1. The scatter plot below shows the height and speed of some of the world's fastest roller coasters. Draw a line that you think is a good fit for the data.

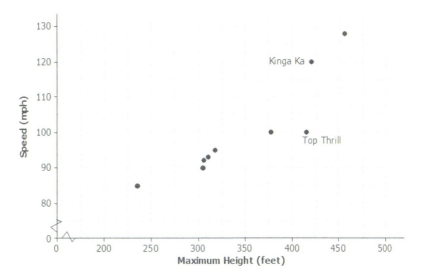

Data Source: http://rcdb.com/rhr.htm

2. Find the equation of your line. Show your steps.

3. For the two roller coasters identified in the scatter plot, use the line to find the approximate difference between the observed speeds and the predicted speeds.

Exit Ticket Sample Solutions

1. The scatter plot below shows the height and speed of some of the world's fastest roller coasters. Draw a line that you think is a good fit for the data.

   Students would draw a line based on the goal of a best fit for the given scatter plot. A possible line is drawn below.

   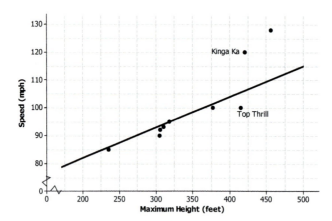

   Data Source: http://rcdb.com/rhr.htm

2. Find the equation of your line. Show your steps.

   Answers will vary based on the line drawn. Let S equal the speed of the roller coaster and H equal the maximum height of the roller coaster.

   $$m = \frac{115 - 85}{500 - 225} \approx 0.11$$

   $$S = 0.11H + b$$
   $$85 = 0.11(225) + b$$
   $$b \approx 60$$

   Therefore, the equation of the line drawn in Problem 1 is $S = 0.11H + 60$.

3. For the two roller coasters identified in the scatter plot, use the line to find the approximate difference between the observed speeds and the predicted speeds.

   Answers will vary depending on the line drawn by a student or the equation of the line. For the Top Thrill, the maximum height is about 415 feet and the speed is about 100 miles per hour. The line indicated in Problem 2 predicts a speed of 106 miles per hour, so the difference is about 6 miles per hour over the actual speed. For the Kinga Ka, the maximum height is about 424 feet with a speed of 120 miles per hour. The line predicts a speed of about 107 miles per hour, for a difference of 13 miles per hour under the actual speed. (Students can use the graph or the equation to find the predicted speed.)

Problem Set Sample Solutions

1. The Monopoly board game is popular in many countries. The scatter plot below shows the distance from "Go" to a property (in number of spaces moving from "Go" in a clockwise direction) and the price of the properties on the Monopoly board. The equation of the line is $P = 8x + 40$, where P represents the price (in Monopoly dollars) and x represents the distance (in number of spaces).

Distance from "Go" (number of spaces)	Price of Property (Monopoly dollars)	Distance from "Go" (number of spaces)	Price of Property (Monopoly dollars)
1	60	21	220
3	60	23	220
5	200	24	240
6	100	25	200
8	100	26	260
9	120	27	260
11	140	28	150
12	150	29	280
13	140	31	300
14	160	32	300
15	200	34	320
16	180	35	200
18	180	37	350
19	200	39	400

Price of Property Versus Distance from "Go" in Monopoly

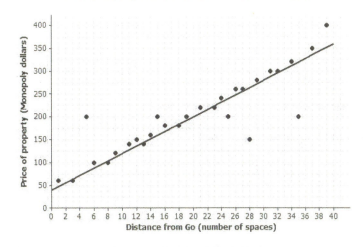

   a. Use the equation to find the difference (observed value−predicted value) for the most expensive property and for the property that is 35 spaces from "Go."

   The most expensive property is 39 spaces from "Go" and costs $400. The price predicted by the line would be $8(39) + 40$, or $352. Observed price − predicted price would be $400 − $352 = $48. The price predicted for 35 spaces from "Go" would be $8(35) + 40$, or $320. Observed price − predicted price would be $200 − $320 = −$120.

   b. Five of the points seem to lie in a horizontal line. What do these points have in common? What is the equation of the line containing those five points?

   These points all have the same price. The equation of the horizontal line through those points would be $P = 200$.

c. Four of the five points described in part (b) are the railroads. If you were fitting a line to predict price with distance from "Go," would you use those four points? Why or why not?

Answers will vary. Because the four points are not part of the overall trend in the price of the properties, I would not use them to determine a line that describes the relationship. I can show this by finding the total error to measure the fit of the line.

2. The table below gives the coordinates of the five points shown in the scatter plots that follow. The scatter plots show two different lines.

Data Point	Independent Variable	Response Variable
A	20	27
B	22	21
C	25	24
D	31	18
E	40	12

Line 1

Line 2

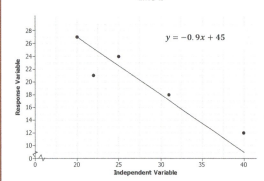
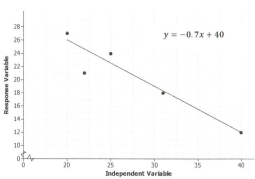

a. Find the predicted response values for each of the two lines.

Independent	Observed Response	Response Predicted by Line 1	Response Predicted by Line 2
20	27	27	26
22	21	25.2	24.6
25	24	22.5	22.5
31	18	17.1	18.3
40	12	9	12

b. For which data points is the prediction based on Line 1 closer to the actual value than the prediction based on Line 2?

It is only for data point A. For data point C, both lines are off by the same amount.

c. Which line (Line 1 or Line 2) would you select as a better fit? Explain.

Line 2 is a better fit because it is closer to more of the data points.

3. The scatter plots below show different lines that students used to model the relationship between body mass (in pounds) and bite force (in pounds) for crocodilians.

   a. Match each graph to one of the equations below, and explain your reasoning. Let B represent bite force (in pounds) and W represent body mass (in pounds).

Equation 1	Equation 2	Equation 3
$B = 3.28W + 126$	$B = 3.04W + 351$	$B = 2.16W + 267$

   Equation: **3**

   The intercept of 267 appears to match the graph, which has the second largest intercept.

   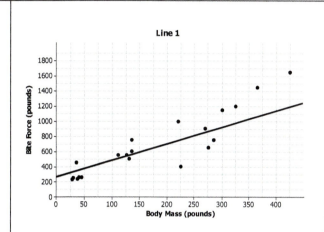

   Equation: **2**

   The intercept of Equation 2 is larger, so it matches Line 2, which has a y-intercept closer to 400.

   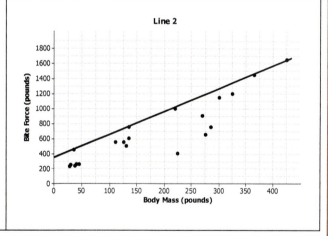

Lesson 9: Determining the Equation of a Line Fit to Data

Equation: 1

The intercept of Equation 1 is the smallest, which seems to match the graph.

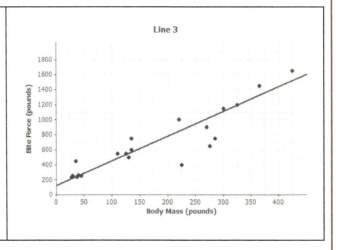

Line 3

b. Which of the lines would best fit the trend in the data? Explain your thinking.

Answers will vary. Line 3 would be better than the other two lines. Line 1 is not a good fit for larger weights, and Line 2 is above nearly all of the points and pretty far away from most of them. It looks like Line 3 would be closer to most of the points.

4. Comment on the following statements:

   a. A line modeling a trend in a scatter plot always goes through the origin.

   Some trend lines go through the origin, but others may not. Often, the value $(0, 0)$ does not make sense for the data.

   b. If the response variable increases as the independent variable decreases, the slope of a line modeling the trend is negative.

   If the trend is from the upper left to the lower right, the slope for the line is negative because for each unit increase in the independent variable, the response decreases.

Name _____ Date _____

1. Many computers come with a Solitaire card game. The player moves cards in certain ways to complete specific patterns. The goal is to finish the game in the shortest number of moves possible, and a player's score is determined by the number of moves. A statistics teacher played the game 16 times and recorded the number of moves and the final score after each game. The line represents the linear function that is used to determine the score from the number of moves.

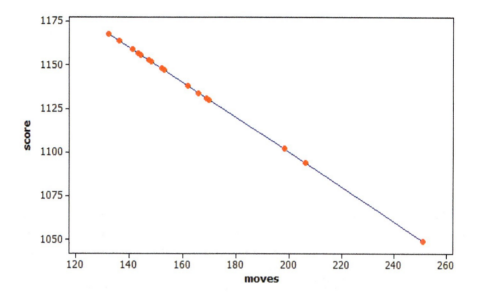

   a. Was this person's average score closer to 1130 or 1110? Explain how you decided.

   b. The first two games she played took 169 moves (1131 points) and 153 moves (1147 points). Based on this information, determine the equation of the linear function used by the computer to calculate the score from the number of moves. Explain your work.

c. Based on the linear function, each time the player makes a move, how many points does she lose?

d. Based on the linear function, how many points does the player start with in this game? Explain your reasoning.

2. To save money, drivers often try to increase their mileage, which is measured in miles per gallon (mpg). One theory is that speed traveled impacts mileage. Suppose the following data are recorded for five different 300-mile tests, with the car traveling at different speeds in miles per hour (mph) for each test.

Speed (mph)	Mileage
50	32
60	29
70	24
80	20
90	17

a. For the data in this table, is the association positive or negative? Explain how you decided.

b. Construct a scatter plot of these data using the following coordinate grid. The vertical axis represents the mileage, and the horizontal axis represents the speed in miles per hour (mph).

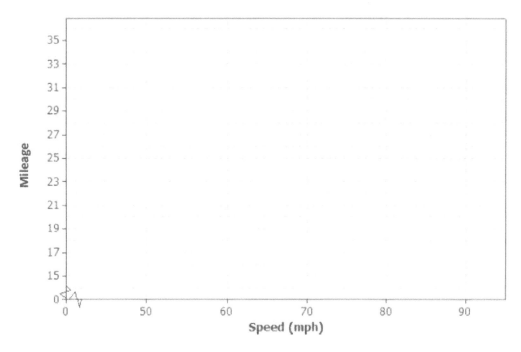

c. Draw a line on your scatter plot that you think is a reasonable model for predicting the mileage from the car speed.

d. Estimate and interpret the slope of the line you found in part (c).

Suppose additional data were measured for three more tests. These results have been added to the previous tests, and the combined data are shown in the table below.

Speed (mph)	Mileage
20	25
30	27
40	30
50	32
60	29
70	24
80	20
90	17

e. Does the association for these data appear to be linear? Why or why not?

f. If your only concern was mileage and you had no traffic constraints, what speed would you recommend traveling based on these data? Explain your choice.

A STORY OF RATIOS   Mid-Module Assessment Task   8•6

A Progression Toward Mastery					
Assessment Task Item		STEP 1 Missing or incorrect answer and little evidence of reasoning or application of mathematics to solve the problem	STEP 2 Missing or incorrect answer but evidence of some reasoning or application of mathematics to solve the problem	STEP 3 A correct answer with some evidence of reasoning or application of mathematics to solve the problem, OR an incorrect answer with substantial evidence of solid reasoning or application of mathematics to solve the problem	STEP 4 A correct answer supported by substantial evidence of solid reasoning or application of mathematics to solve the problem
1	a	Student makes no use of the given data.	Student chooses 1110 based solely on it being the midpoint of the y-axis values.	Student chooses 1130, but reasoning is incomplete or missing.	Student chooses 1130 based on the higher concentration of red dots around those y-values.
	b	Student cannot obtain a line.	Student attempts to estimate a line from the graph.	Student uses a reasonable approach but does not obtain the correct line (e.g., interchanges slope and intercept in the equation, sets up an inverse of the slope equation, or shows insufficient work).	Student finds the correct equation (or with minor errors) from slope $= \frac{(1131-1147)}{169-153} = -1$, and intercept from $1131 = a - 169$, so $a = 1300$. Equation: $y = 1300 - x$, where y represents points and x represents number of moves.
	c	Student makes no use of the given data.	Student does not recognize this as a question about slope.	Student estimates the slope from the graph.	Student reports the slope (-1) found in part (b).
	d	Student makes no use of the given data.	Student does not recognize this as a question about intercept.	Student estimates the intercept from the graph or solves the equation with $x = 0$ without recognizing a connection to the equation.	Student reports the intercept (1300) found in part (b).

Module 6:   Linear Functions

2	a	Student makes no use of the given data.	Student bases the answer solely on the content (e.g., faster cars are less fuel efficient).	Student refers to the scatter plot in part (b) or makes a minor error (e.g., misspeaks and describes a negative association but appears to unintentionally call it a positive association).	Student notes that mileage values are decreasing while speeds (mph) are increasing and states that this is a negative association. OR Student solves for the slope and notes the sign of the slope.
	b	Student makes no use of the given data.	Student does not construct a scatter plot with the correct number of dots.	Student constructs a scatter plot but reverses the roles of speed and mileage.	Student constructs a scatter plot that has five dots in the correct locations.
	c	Student does not answer the question.	Student does not draw a line but rather connects the dots.	Student draws a line that does not reasonably describe the behavior of the plotted data.	Student draws a line that reasonably describes the behavior of the plotted data.
	d	Student makes no use of the given data.	Student uses the correct approach but makes major calculation errors (e.g., using only values from the table or failing to interpret the slope).	Student uses the correct approach but makes minor errors in calculation or in interpretation.	Student estimates the coordinates for two locations and determines the change in y-values divided by the change in x-values, for example, $(50, 33)$ and $(80, 20)$, which yields $\left(-\frac{13}{30}\right) \approx -0.43\overline{3}$, and interprets this as the decrease in mileage per additional mph in speed.
	e	Student does not comment on the increasing or decreasing pattern in the values.	Student attempts to sketch a graph of the data and comments on the overall pattern but does not comment on the change in the direction of the association.	Student comments only on how the change in the mileage is not constant without commenting on the change in the sign of the differences.	Student comments on the increasing and then decreasing behavior of the mileage column as the mileage column steadily increases.
	f	Student does not answer the question.	Student recommends 55 mph based only on anecdote and does not provide any reasoning.	Student recommends a reasonable speed but does not fully justify the choice.	Student recommends and gives justification for a speed between 40 and 50 mph, or at 50 mph, based on the association "peaking" at 50 mph.

Module 6: Linear Functions

Name _____ Date _____

1. Many computers come with a Solitaire card game. The player moves cards in certain ways to complete specific patterns. The goal is to finish the game in the shortest number of moves possible, and a player's score is determined by the number of moves. A statistics teacher played the game 16 times and recorded the number of moves and the final score after each game. The line represents the linear function that is used to determine the score from the number of moves.

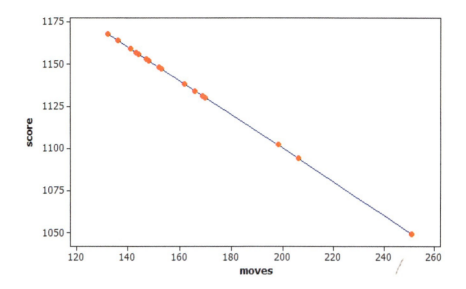

   a. Was this person's average score closer to 1130 or 1110? Explain how you decided.

   Most of the games had scores between 1125 and 1175. The mean score will be closer to 1130.

   b. The first two games she played took 169 moves (1131 points) and 153 moves (1147 points). Based on this information, determine the equation of the linear function used by the computer to calculate the score from the number of moves. Explain your work.

   The difference in the scores is 1131 − 1147 or −16.
   The difference in the number of moves is 169 − 153 = 16
   The slope is −16/16 or −1. This means that
   1131 = intercept − 169, so intercept equals 1300
   Score = 1300 − moves

c. Based on the linear function, each time the player makes a move, how many points does she lose?

One point lost per move.

d. Based on the linear function, how many points does the player start with in this game? Explain your reasoning.

1300, or the score when the number of moves equals 0.

2. To save money, drivers often try to increase their mileage, which is measured in miles per gallon (mpg). One theory is that speed traveled impacts mileage. Suppose the following data are recorded for five different 300-mile tests, with the car traveling at different speeds in miles per hour (mph) for each test.

Speed (mph)	Mileage
50	32
60	29
70	24
80	20
90	17

a. For the data in this table, is the association positive or negative? Explain how you decided.

As the speed increases in miles per hour, the miles per gallon decrease. This describes a negative association.

b. Construct a scatter plot of these data using the following coordinate grid. The vertical axis represents the mileage and the horizontal axis represents the speed in miles per hour (mph).

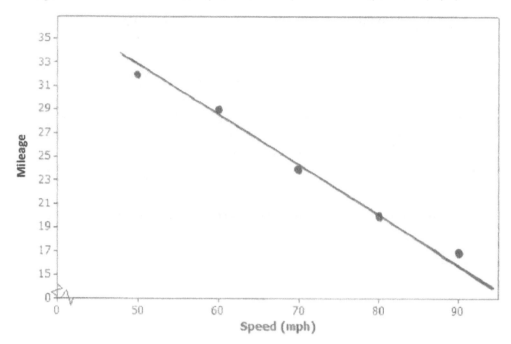

c. Draw a line on your scatter plot that you think is a reasonable model for predicting the mileage from the car speed.

d. Estimate and interpret the slope of the line you found in part (c).

Two points are approximately $(80, 20)$ and $(50, 33)$.

So, slope $\approx \dfrac{20 - 33}{80 - 50} \approx -0.4\overline{33}$

Each increase of 1 mph in speed predicts a decrease of $0.4\overline{33}$ mpg.

Suppose additional data were measured for three more tests. These results have been added to the previous tests, and the combined data are shown in the table below.

Speed (mph)	Mileage
20	25
30	27
40	30
50	32
60	29
70	24
80	20
90	17

e. Does the association for these data appear to be linear? Why or why not?

No, while the speeds increase, the mileage values increase and then mostly decrease. There is no fixed rate of increase or decrease for mileage based on increased speed.

f. If your only concern was mileage and you had no traffic constraints, what speed would you recommend traveling based on these data? Explain your choice.

About 50 mph. It is around 50 mph that the mpg stops increasing and starts to decrease.

A STORY OF RATIOS

GRADE 8

Mathematics Curriculum

GRADE 8 • MODULE 6

Topic C
Linear and Nonlinear Models

Focus Standards:	■ Construct and interpret scatter plots for bivariate measurement data to investigate patterns of association between two quantities. Describe patterns such as clustering, outliers, positive or negative association, linear association, and nonlinear association.
	■ Know that straight lines are widely used to model relationships between two quantitative variables. For scatter plots that suggest a linear association, informally fit a straight line, and informally assess the model fit by judging the closeness of the data points to the line.
	■ Use the equation of a linear model to solve problems in the context of bivariate measurement data, interpreting the slope and intercept. *For example, in a linear model for a biology experiment, interpret a slope of* 1.5 cm/hr *as meaning that an additional hour of sunlight each day is associated with an additional* 1.5 cm *in mature plant height.*
Instructional Days:	3
Lesson 10:	Linear Models (P)[1]
Lesson 11:	Using Linear Models in a Data Context (P)
Lesson 12:	Nonlinear Models in a Data Context (Optional) (P)

In Topic C, students interpret and use linear models. They provide verbal descriptions based on how one variable changes as the other variable changes. Students identify and describe how one variable changes as the other variable changes for linear and nonlinear associations. They describe patterns of positive and negative associations using scatter plots. In Lesson 10, students identify applications in which a linear function models the relationship between two numerical variables. In Lesson 11, students use a linear model to answer questions about the relationship between two numerical variables by interpreting the context of a data set. Students use graphs and the patterns of linear association to answer questions about the relationship of the data. In Lesson 12, students also examine patterns and graphs that describe nonlinear associations of data.

---

[1]Lesson Structure Key: **P**-Problem Set Lesson, **M**-Modeling Cycle Lesson, **E**-Exploration Lesson, **S**-Socratic Lesson

Topic C: Linear and Nonlinear Models

Lesson 10: Linear Models

Student Outcomes

- Students identify situations where it is reasonable to use a linear function to model the relationship between two numerical variables.
- Students interpret slope and the initial value in a data context.

Lesson Notes

In previous lessons, students were given a set of bivariate data on variables that were linearly related. Students constructed a scatter plot of the data, informally fit a line to the data, and found the equation of their prediction line. The lessons also discussed criteria students could use to determine what might be considered the *best-fitting* prediction line for a given set of data. A more formal discussion of this topic occurs in Algebra I.

This lesson introduces a formal statistical terminology for the two variables that define a bivariate data set. In a prediction context, the x-variable is referred to as the *independent variable, explanatory variable,* or *predictor variable*. The y-variable is referred to as the *dependent variable, response variable,* or *predicted variable*. Students should become equally comfortable with using the pairings (independent, dependent), (explanatory, response), and (predictor, predicted). Statistics builds on data, and in this lesson, students investigate bivariate data that are linearly related. Students examine how the dependent variable relates to the independent variable or how the predicted variable relates to the predictor variable. Students also need to connect the linear function in words to a symbolic form that represents a linear function. In most cases, the independent variable is denoted by x and the dependent variable by y.

Similar to lessons at the beginning of this module, this lesson works with *exact* linear relationships. This is done to build conceptual understanding of how structural elements of the modeling equation are explained in context. Students apply this thinking to more authentic data contexts in the next lesson.

Classwork

In previous lessons, you used data that follow a linear trend either in the positive direction or the negative direction and informally fit a line through the data. You determined the equation of an informal fitted line and used it to make predictions.

In this lesson, you use a function to model a linear relationship between two numerical variables and interpret the slope and intercept of the linear model in the context of the data. Recall that a function is a rule that relates a dependent variable to an independent variable.

In statistics, a dependent variable is also called a *response variable* or a *predicted variable*. An independent variable is also called an *explanatory variable* or a *predictor variable*.

Scaffolding:
- A dependent variable is also called a *response* or *predicted* variable.
- An independent variable is also called an *explanatory* or a *predictor* variable.
- It is important to make the interchangeability of these terms clear to English language learners.
- For each of the pairings, students should have the chance to read, write, speak, and hear them on multiple occasions.

A STORY OF RATIOS   Lesson 10   8•6

Example 1 (5 minutes)

This lesson begins by challenging students' understanding of the terminology. Read through the opening text, and explain the difference between dependent and independent variables. Pose the question to the class at the end of the example, and allow for multiple responses.

- What are some other possible numerical independent variables that could relate to how well you are going to do on the quiz?
    - *How many hours of sleep I got the night before*

> **Example 1**
>
> Predicting the value of a numerical dependent (response) variable based on the value of a given numerical independent variable has many applications in statistics. The first step in the process is to identify the dependent (predicted) variable and the independent (predictor) variable.
>
> There may be several independent variables that might be used to predict a given dependent variable. For example, suppose you want to predict how well you are going to do on an upcoming statistics quiz. One possible independent variable is how much time you spent studying for the quiz. What are some other possible numerical independent variables that could relate to how well you are going to do on the quiz?

Exercise 1 (5 minutes)

Exercise 1 requires students to write two possible explanatory variables that might be used for each of several given response variables. Give students a moment to think about each response variable, and then discuss the answers as a class. Allow for multiple student responses.

> **Exercises 1–2**
>
> 1. For each of the following dependent (response) variables, identify two possible numerical independent (explanatory) variables that might be used to predict the value of the dependent variable.
>
>   *Answers will vary. Here again, make sure that students are defining their explanatory variables (predictors) correctly and that they are numerical.*
>
> | Response Variable | Possible Explanatory Variables |
> |---|---|
> | Height of a son | 1. Height of the boy's father
2. Height of the boy's mother |
> | Number of points scored in a game by a basketball player | 1. Number of shots taken in the game
2. Number of minutes played in the game |
> | Number of hamburgers to make for a family picnic | 1. Number of people in the family
2. Price of hamburger meat |
> | Time it takes a person to run a mile | 1. Height above sea level of the track field
2. Number of practice days |
> | Amount of money won by a contestant on Jeopardy!™ (television game show) | 1. IQ of the contestant
2. Number of questions correctly answered |
> | Fuel efficiency (in miles per gallon) for a car | 1. Weight of the car
2. Size of the car's engine |
> | Number of honey bees in a beehive at a particular time | 1. Size of a queen bee
2. Amount of honey harvested from the hive |
> | Number of blooms on a dahlia plant | 1. Amount of fertilizer applied to the plant
2. Amount of water applied to the plant |
> | Number of forest fires in a state during a particular year | 1. Number of acres of forest in the state
2. Amount of rain in the state that year |

Lesson 10:   Linear Models

A STORY OF RATIOS   Lesson 10   8•6

Exercise 2 (5 minutes)

This exercise reverses the format and asks students to provide a response variable for each of several given explanatory variables. Again, give students a moment to consider each independent variable. Then, discuss the dependent variables as a class. Allow for multiple student responses.

2. Now, reverse your thinking. For each of the following numerical independent variables, write a possible numerical dependent variable.

Dependent Variable	Possible Independent Variables
Time it takes a student to run a mile	Age of a student
Distance a golfer drives a ball from a tee	Height of a golfer
Time it takes pain to disappear	Amount of a pain reliever taken
Amount of money a person makes in a lifetime	Number of years of education
Number of tomatoes harvested in a season	Amount of fertilizer used on a garden
Price of a diamond ring	Size of a diamond in a ring
A baseball team's batting average	Total salary for all of a team's players

Example 2 (3–5 minutes)

This example begins the study of an exact linear relationship between two numerical variables. Example 2 and Exercises 3–9 address bivariate data that have an exact functional form, namely, linear. Students become familiar with an equation of the form $y = \text{intercept} + (\text{slope})x$. They connect this representation to the equation of a linear function ($y = mx + b$ or $y = a + bx$) developed in previous modules. Make sure students clearly identify the slope and the y-intercept as they describe a linear function. Students interpret slope as the change in the dependent variable (the y-variable) for an increase of one unit in the independent variable (the x-variable).

For example, if exam score $= 57 + 8$ (study time), or equivalently, $y = 57 + 8x$, where y represents the exam score and x represents the study time in hours, then an increase of one hour in study time produces an increase of 8 points in the predicted exam score. Encourage students to interpret slope in the context of the problem. Their interpretation of slope as simply "rise over run" is not sufficient in a statistical setting.

Students should become comfortable writing linear models using descriptive words (such as *exam score* and *study time*) or using symbols, such as x and y, to represent variables. Using descriptive words when writing model equations can help students keep the context in mind, which is important in statistics.

Note that bivariate numerical data that do not have an exact linear functional form but do have a linear trend are covered in the next lesson. Starting with Example 2, this lesson covers only contexts in which the linear relationship is exact.

Give students a moment to read through Example 2. For English language learners, consider reading the example aloud.

Example 2

A cell phone company offers the following basic cell phone plan to its customers: A customer pays a monthly fee of 40.00. In addition, the customer pays 0.15 per text message sent from the cell phone. There is no limit to the number of text messages per month that could be sent, and there is no charge for receiving text messages.

Lesson 10:   Linear Models

Lesson 10

Exercises 3–9 (10–15 minutes)

These exercises build on earlier lessons in Module 6. Provide time for students to develop answers to the exercises. Then, confirm their answers as a class.

Exercises 3–11

3. Determine the following:

   a. Justin never sends a text message. What would be his total monthly cost?

      Justin's monthly cost would be $\$40.00$.

   b. During a typical month, Abbey sends 25 text messages. What is her total cost for a typical month?

      Abbey's monthly cost would be $\$40.00 + \$0.15(25)$, *or* $\$43.75$.

   c. Robert sends at least 250 text messages a month. What would be an estimate of the least his total monthly cost is likely to be?

      Robert's monthly cost would be $\$40.00 + \$0.15(250)$, *or* $\$77.50$.

   Scaffolding:

   Using a table may help students better understand the relationship between the number of text messages and total monthly cost.

Number of messages	Total Cost ($)
0	$40 + 0.15(0) = 40$
1	$40 + 0.15(1) = 40.15$
2	$40 + 0.15(2) = 40.30$
3	$40 + 0.15(3) = 40.45$
4	$40 + 0.15(4) = 40.60$
5	$40 + 0.15(5) = 40.75$
10	$40 + 0.15(10) = 41.50$

   There is a cost increase of $0.15 for every additional text message sent from the phone.

4. Use descriptive words to write a linear model describing the relationship between the number of text messages sent and the total monthly cost.

   Total monthly cost $= \$40.00 +$ *(number of text messages)* $\cdot \$0.15$

5. Is the relationship between the number of text messages sent and the total monthly cost linear? Explain your answer.

   Yes. For each text message, the total monthly cost goes up by $\$0.15$. *From our previous work with linear functions, this would indicate a linear relationship.*

6. Let x represent the independent variable and y represent the dependent variable. Use the variables x and y to write the function representing the relationship you indicated in Exercise 4.

   Students show the process in developing a model of the relationship between the two variables.
   $y = 0.15x + 40$ *or* $y = 40 + 0.15x$

7. Explain what $\$0.15$ represents in this relationship.

   $\$0.15$ *represents the slope of the linear relationship, or the change in the total monthly cost is* $\$0.15$ *for an increase of one text message.* (Students need to clearly explain that slope is the change in the dependent variable for a 1-unit increase in the independent variable.)

8. Explain what $\$40.00$ represents in this relationship.

   $\$40.00$ *represents the fixed monthly fee or the y-intercept of this relationship. This is the value of the total monthly cost when the number of text messages is* 0.

Lesson 10: Linear Models

9. Sketch a graph of this relationship on the following coordinate grid. Clearly label the axes, and include units in the labels.

Anticipated response: Students label the x-axis as the number of text messages. They label the y-axis as the total monthly cost. Students use any two points they derived in Exercise 3. The following graph uses the point of $(0, 40)$ for Justin and the point $(250, 77.5)$ for Robert. Highlight the intercept of $(0, 40)$, along with the slope of the line they sketched. Also, point out that the line students draw should be a dotted line (and not a solid line). The number of text messages can only be whole numbers, and as a result, the line representing this relationship should indicate that values in between the whole numbers representing the text messages are not part of the data.

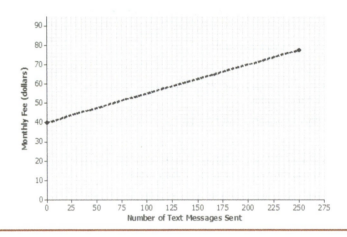

Exercise 10 (5 minutes)

If time is running short, teachers may want to choose either Exercise 10 or 11 to develop in class and assign the other to the Problem Set. Let students continue to work with a partner, and confirm answers as a class.

10. LaMoyne needs four more pieces of lumber for his Scout project. The pieces can be cut from one large piece of lumber according to the following pattern.

The lumberyard will make the cuts for LaMoyne at a fixed cost of $2.25 plus an additional cost of 25 cents per cut. One cut is free.

a. What is the functional relationship between the total cost of cutting a piece of lumber and the number of cuts required? What is the equation of this function? Be sure to define the variables in the context of this problem.

As students uncover the information in this problem, they should realize that the functional relationship between the total cost and number of cuts is linear. Noting that one cut is free, the equation could be written in one of the following ways:

Total cost for cutting $= 2.25 + (0.25)$(number of cuts $- 1$)
$y = 2.25 + (0.25)(x - 1)$, where x is the number of cuts and y is the total cost for cutting

Total cost for cutting $= 2 + (0.25)$(number of cuts)
$y = 2 + 0.25x$, where x is the number of cuts and y is the total cost for cutting

Total cost for cutting $= 2.25 + (0.25)$(number of paid cuts)
$y = 2.25 + 0.25x$, where x is the number of paid cuts and y is the total cost for cutting

b. Use the equation to determine LaMoyne's total cost for cutting.

LaMoyne requires three cuts, one of which is free. Using any of the three forms given in part (a) yields a total cost for cutting of $2.75.

c. In the context of this problem, interpret the slope of the equation in words.

Using any of the three forms, each additional cut beyond the free one adds $0.25 to the total cost for cutting.

d. Interpret the y-intercept of your equation in words in the context of this problem. Does interpreting the intercept make sense in this problem? Explain.

If no cuts are required, then there is no fixed cost for cutting. So, it does not make sense to interpret the intercept in the context of this problem.

Exercise 11 (5–7 minutes)

Let students work with a partner. Then, confirm answers as a class.

11. Omar and Olivia were curious about the size of coins. They measured the diameter and circumference of several coins and found the following data.

U.S. Coin	Diameter (millimeters)	Circumference (millimeters)
Penny	19.0	59.7
Nickel	21.2	66.6
Dime	17.9	56.2
Quarter	24.3	76.3
Half Dollar	30.6	96.1

a. Wondering if there was any relationship between diameter and circumference, they thought about drawing a picture. Draw a scatter plot that displays circumference in terms of diameter.

Students may need some help in deciding which is the independent variable and which is the dependent variable. Hopefully, they have seen from previous problems that whenever one variable, say variable A, is to be expressed in terms of some variable B, then variable A is the dependent variable, and variable B is the independent variable. So, circumference is being taken as the dependent variable in this problem, and diameter is being taken as the independent variable.

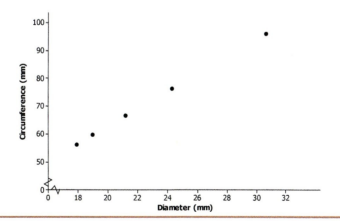

Lesson 10: Linear Models

> b. Do you think that circumference and diameter are related? Explain.
>
> *It may be necessary to point out to students that because the data are rounded to one decimal place, the points on the scatter plot may not fall exactly on a line; however, they should. Circumference and diameter are linearly related.*
>
> c. Find the equation of the function relating circumference to the diameter of a coin.
>
> *Again, because of a rounding error, equations that students find may be slightly different depending on which points they choose to do their calculations. Hopefully, they all arrive at something close to a circumference equal to 3.14, or π, multiplied by diameter.*
>
> *For example, the slope of the line containing points $(19, 59.7)$ and $(30.6, 96.1)$ is $\frac{96.1 - 59.7}{30.6 - 19} = 3.1379$, which rounds to 3.14.*
>
> *The intercept may be found using $59.7 = a + (3.14)(19.0)$, which yields $a = 0.04$, which rounds to 0.*
>
> *Therefore, $C = 3.14d + 0 = 3.14d$.*
>
> d. The value of the slope is approximately equal to the value of π. Explain why this makes sense.
>
> *The slope is identified as being approximately equal to π. (Note: Most students have previously studied the relationship between circumference and diameter of a circle. However, if students have not yet seen this result, discuss the interesting result that if the circumference of a circle is divided by its diameter, the result is a constant, namely, 3.14 rounded to two decimal places, no matter what circle is being considered.)*
>
> e. What is the value of the y-intercept? Explain why this makes sense.
>
> *If the diameter of a circle is 0 (a point), then according to the equation, its circumference is 0. That is true, so interpreting the intercept of 0 makes sense in this problem.*

Closing (2–3 minutes)

- Think back to Exercise 10. If the equation that models LaMoyne's total cost of cutting is given by $y = 2.25 + 0.25x$, what are the dependent and independent variables?
    - *The independent variable is the number of paid cuts. The dependent variable is the total cost for cutting.*
- What are the meanings of the y-intercept and slope in context?
    - *The y-intercept is the fee for the first cut; however, if no cuts are required, then there is no fixed cost for cutting. The slope is the cost per cut after the first.*
- How are these examples different from the data we have been studying before this lesson?
    - *These examples are exact linear relationships.*

Lesson Summary

- A linear functional relationship between a dependent and an independent numerical variable has the form $y = mx + b$ or $y = a + bx$.
- In statistics, a dependent variable is one that is predicted, and an independent variable is the one that is used to make the prediction.
- The graph of a linear function describing the relationship between two variables is a line.

Exit Ticket (5–7 minutes)

Lesson 10: Linear Models

Lesson 10: Linear Models

Exit Ticket

Suppose that a cell phone monthly rate plan costs the user 5 cents per minute beyond a fixed monthly fee of $20. This implies that the relationship between monthly cost and monthly number of minutes is linear.

1. Write an equation in words that relates total monthly cost to monthly minutes used. Explain how you found your answer.

2. Write an equation in symbols that relates the total monthly cost in dollars (y) to monthly minutes used (x).

3. What is the cost for a month in which 182 minutes are used? Express your answer in words in the context of this problem.

Exit Ticket Sample Solutions

> Suppose that a cell phone monthly rate plan costs the user 5 cents per minute beyond a fixed monthly fee of $20. This implies that the relationship between monthly cost and monthly number of minutes is linear.
>
> 1. Write an equation in words that relates total monthly cost to monthly minutes used. Explain how you found your answer.
>
>   *The equation is given by* total monthly cost $= 20 + 0.05$ (number of minutes used for a month).
>
>   *The y-intercept in the equation is the fixed monthly cost,* $20.
>
>   *The slope is the amount paid per minute of cell phone usage, or* $0.05 *per minute.*
>
>   *The linear form is* total monthly cost $=$ fixed cost $+$
>   cost per minute (number of minutes used for a month).
>
> 2. Write an equation in symbols that relates the total monthly cost in dollars (y) to monthly minutes used (x).
>
>   *The equation is* $y = 20 + 0.05x$, *where y is the total cost for a month in dollars and x is cell phone usage for the month in minutes.*
>
> 3. What is the cost for a month in which 182 minutes are used? Express your answer in words in the context of this problem.
>
> $20 + (0.05)(182) = 29.10$
>
>   *The total monthly cost in a month using* 182 *minutes would be* $29.10.

Problem Set Sample Solutions

> 1. The Mathematics Club at your school is having a meeting. The advisor decides to bring bagels and his award-winning strawberry cream cheese. To determine his cost, from past experience he figures 1.5 bagels per student. A bagel costs 65 cents, and the special cream cheese costs $3.85 and will be able to serve all of the anticipated students attending the meeting.
>
>   a. Find an equation that relates his total cost to the number of students he thinks will attend the meeting.
>
>   *Encourage students to write a problem in words in its context. For example, the advisor's total cost $=$ cream cheese fixed cost $+$ cost of bagels. The cost of bagels depends on the unit cost of a bagel times the number of bagels per student times the number of students. So, with symbols, if c denotes the total cost in dollars and n denotes the number of students, then $c = 3.85 + (0.65)(1.5)(n)$, or $c = 3.85 + 0.975n$.*
>
>   b. In the context of the problem, interpret the slope of the equation in words.
>
>   *For each additional student, the cost goes up by* 0.975 *dollar, or* 97.5 *cents.*
>
>   c. In the context of the problem, interpret the y-intercept of the equation in words. Does interpreting the intercept make sense? Explain.
>
>   *If there are no students, the total cost is* $3.85. *Students could interpret this by saying that the meeting was called off before any bagels were bought, but the advisor had already made his award-winning cream cheese, so the cost is* $3.85. *The intercept makes sense. Other students might argue otherwise.*

Lesson 10: Linear Models

2. John, Dawn, and Ron agree to walk/jog for 45 minutes. John has arthritic knees but manages to walk $1\frac{1}{2}$ miles. Dawn walks $2\frac{1}{4}$ miles, while Ron manages to jog 6 miles.

   a. Draw an appropriate graph, and connect the points to show that there is a linear relationship between the distance that each traveled based on how fast each traveled (speed). Note that the speed for a person who travels 3 miles in 45 minutes, or $\frac{3}{4}$ hour, is found using the expression $3 \div \frac{3}{4}$, which is 4 miles per hour.

   John's speed is 2 miles per hour because $1\frac{1}{2} \div \frac{3}{4} = 2$. Dawn's speed is 3 miles per hour because $2\frac{1}{4} \div \frac{3}{4} = 3$. Ron's speed is 8 miles per hour because $6 \div \frac{3}{4} = 8$. Students may draw the scatter plot incorrectly. Note that distance is to be expressed in terms of speed so that distance is the dependent variable on the vertical axis, and speed is the independent variable on the horizontal axis.

   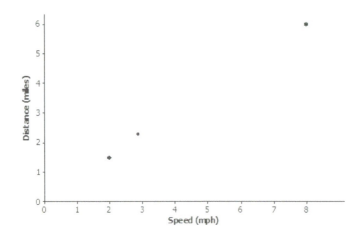

   b. Find an equation that expresses distance in terms of speed (how fast one goes).

   The slope is $\frac{6 - 1.5}{8 - 2} = 0.75$, so the equation of the line through these points is distance $= a + (0.75)$(speed).

   Next, find the intercept. For example, solve $6 = a + (0.75)(8)$ for a, which yields $a = 0$.

   So, the equation is distance $= 0.75$(speed).

   c. In the context of the problem, interpret the slope of the equation in words.

   If someone increases her speed by 1 mile per hour, then that person travels 0.75 additional mile in 45 minutes.

   d. In the context of the problem, interpret the y-intercept of the equation in words. Does interpreting the intercept make sense? Explain.

   The intercept of 0 makes sense because if the speed is 0 miles per hour, then the person is not moving. So, the person travels no distance.

Note: Simple interest is developed in the next problem. It is an excellent example of an application of a linear function. If students have not worked previously with finance problems of this type, it may be necessary to carefully explain simple interest as stated in the problem. It is an important discussion to have with students if time permits. If this discussion is not possible and students have not worked previously with any financial applications, then omit this problem.

3. Simple interest is money that is paid on a loan. Simple interest is calculated by taking the amount of the loan and multiplying it by the rate of interest per year and the number of years the loan is outstanding. For college, Jodie's older brother has taken out a student loan for $4,500 at an annual interest rate of 5.6%, or 0.056. When he graduates in four years, he has to pay back the loan amount plus interest for four years. Jodie is curious as to how much her brother has to pay.

   a. Jodie claims that her brother has to pay a total of $5,508. Do you agree? Explain. As an example, a $1,200 loan has an 8% annual interest rate. The simple interest for one year is $96 because $(0.08)(1,200) = 96$. The simple interest for two years would be $192 because $(2)(96) = 192$.

   The total cost to repay = amount of the loan + interest on the loan.

   Interest on the loan is the annual interest times the number of years the loan is outstanding.

   The annual interest amount is $(0.056)(\$4,500) = \$252$.

   For four years, the simple interest amount is $4(\$252) = \$1,008$.

   So, the total cost to repay the loan is $\$4,500 + \$1,008 = \$5,508$. Jodie is right.

   b. Write an equation for the total cost to repay a loan of P if the rate of interest for a year is r (expressed as a decimal) for a time span of t years.

   Note: Work with students in identifying variables to represent the values discussed in this exercise. For example, the total cost to repay a loan is the amount of the loan plus the simple interest, or $P + I$, where P represents the amount of the loan and I represents the simple interest over t years.

   The amount of interest per year is P times the annual interest. Let r represent the interest rate per year as a decimal.

   The amount of interest per year is the amount of the loan, P, multiplied by the annual interest rate as a decimal, r (e.g., 5% is 0.05). The simple interest for t years, I, is the amount of interest per year multiplied by the number of years: $I = (rt)P$.

   The total cost to repay the loan, c, is the amount of the loan plus the amount of simple interest; therefore, $c = P + (rt)P$.

   c. If P and r are known, is the equation a linear equation?

   If P and r are known, then the equation should be written as $c = P + (rP)t$, which is the linear form where c is the dependent variable and t is the independent variable.

   d. In the context of this problem, interpret the slope of the equation in words.

   For each additional year that the loan is outstanding, the total cost to repay the loan is increased by $\$rP$.

   As an example, consider Jodie's brother's equation for t years: $c = 4500 + (0.056)(4500)t$, or $c = 4500 + 252t$. For each additional year that the loan is not paid off, the total cost increases by $\$252$.

   e. In the context of this problem, interpret the y-intercept of the equation in words. Does interpreting the intercept make sense? Explain.

   The y-intercept is the value where $t = 0$. In this problem, it is the cost of the loan at the time that the loan was taken out. This makes sense because after 0 years, the cost to repay the loan would be $\$4,500$, which is the amount of the original loan.

Lesson 10: Linear Models

Lesson 11: Using Linear Models in a Data Context

Student Outcomes

- Students recognize and justify that a linear model can be used to fit data.
- Students interpret the slope of a linear model to answer questions or to solve a problem.

Lesson Notes

In a previous lesson, students were given bivariate numerical data where there was an exact linear relationship between two variables. Students identified which variable was the predictor variable (i.e., independent variable) and which was the predicted variable (i.e., dependent variable). They found the equation of the line that fit the data and interpreted the intercept and slope in words in the context of the problem. Students also calculated a prediction for a given value of the predictor variable. This lesson introduces students to data that are not exactly linear but that have a linear trend. Students informally fit a line and use it to make predictions and answer questions in context.

Although students may want to rely on using symbolic representations for lines, it is important to challenge them to express their equations in words in the context of the problem. Keep emphasizing the meaning of slope in context, and avoid the use of "rise over run." Slope is the impact that increasing the value of the predictor variable by one unit has on the predicted value.

Classwork

Exercise 1 (10–12 minutes)

Introduce the data in the exercise. Using a short video may help students (especially English language learners) to better understand the context of the data. Then, work through each part of the exercise as a class. Ask students the following:

- Looking at the table, what trend appears in the data?
    - *There is a positive trend. As one variable increases in value, so does the other.*
- Looking at the scatter plots, is there an exact linear relationship between the variables?
    - *No, the four points cannot be connected by a straight line.*

Exercises

1. Old Faithful is a geyser in Yellowstone National Park. The following table offers some rough estimates of the length of an eruption (in minutes) and the amount of water (in gallons) in that eruption.

Length (minutes)	1.5	2	3	4.5
Amount of Water (gallons)	3,700	4,100	6,450	8,400

This data is consistent with actual eruption and summary statistics that can be found at the following links:
http://geysertimes.org/geyser.php?id=OldFaithful and http://www.yellowstonepark.com/2011/07/about-old-faithful/

a. Chang wants to predict the amount of water in an eruption based on the length of the eruption. What should he use as the dependent variable? Why?

Since Chang wants to predict the amount of water in an eruption, the time length (in minutes) is the predictor, and the amount of water is the dependent variable.

> **Scaffolding:**
> Make the interchangeability of the terms *linearly related* and *linear relationship* clear to students.

b. Which of the following two scatter plots should Chang use to build his prediction model? Explain.

The predicted variable goes on the vertical axis with the predictor on the horizontal axis. So, the amount of water goes on the y-axis. The plot on the graph on the right should be used.

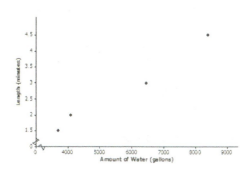
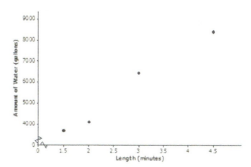

c. Suppose that Chang believes the variables to be linearly related. Use the *first* and *last* data points in the table to create a linear prediction model.

$$m = \frac{8400 - 3700}{4.5 - 1.5} \approx 1,566.7$$

So, $y = a + (1,566.7)x$.

Using either $(1.5, 3700)$ or $(4.5, 8400)$ allows students to solve for the intercept. For example, solving $3,700 = a + (1,566.7)(1.5)$ for a yields $a = 1,349.95$, or rounded to $1,350.0$ gallons. Be sure students talk through the units in each step of the calculations.

The (informal) linear prediction model is $y = 1,350.0 + 1,566.7x$. The amount of water (y) is in gallons, and the length of the eruption (x) is in minutes.

d. A friend of Chang's told him that Old Faithful produces about $3,000$ gallons of water for every minute that it erupts. Does the linear model from part (c) support what Chang's friend said? Explain.

This question requires students to interpret slope. An additional minute in eruption length results in a prediction of an additional $1,566.7$ gallons of water produced. So, Chang's friend who claims Old Faithful produces $3,000$ gallons of water a minute must be thinking of a different geyser.

e. Using the linear model from part (c), does it make sense to interpret the y-intercept in the context of this problem? Explain.

No, it doesn't make sense because if the length of an eruption is 0, then it cannot produce $1,350$ gallons of water. (Convey to students that some linear models have y-intercepts that do not make sense within the context of a problem.)

A STORY OF RATIOS   Lesson 11 8•6

Exercise 2 (15–20 minutes)

Let students work in small groups or with a partner. Introduce the data in the table. Note that the mean times of the three medal winners are provided for each year. Let students work on the exercise, and confirm answers to parts (c)–(f) as a class. After answers have been confirmed, ask the class:

- What is the meaning of the y-intercept from part (c)?
    - The y-intercept from part (c) is $(0, 34.91)$. It does not make sense within the context of the problem. In Year 0, the mean medal time was 34.91 seconds.

2. The following table gives the times of the gold, silver, and bronze medal winners for the men's 100-meter race (in seconds) for the past 10 Olympic Games.

Year	2012	2008	2004	2000	1996	1992	1988	1984	1980	1976
Gold	9.63	9.69	9.85	9.87	9.84	9.96	9.92	9.99	10.25	10.06
Silver	9.75	9.89	9.86	9.99	9.89	10.02	9.97	10.19	10.25	10.07
Bronze	9.79	9.91	9.87	10.04	9.90	10.04	9.99	10.22	10.39	10.14
Mean Time	9.72	9.83	9.86	9.97	9.88	10.01	9.96	10.13	10.30	10.09

Data Source: https://en.wikipedia.org/wiki/100_metres_at_the_Olympics#Men

a. If you wanted to describe how mean times change over the years, which variable would you use as the independent variable, and which would you use as the dependent variable?

Mean medal time (dependent variable) is being predicted based on year (independent variable).

b. Draw a scatter plot to determine if the relationship between mean time and year appears to be linear. Comment on any trend or pattern that you see in the scatter plot.

The scatter plot indicates a negative trend, meaning that, in general, the mean race times have been decreasing over the years even though there is not a perfect linear pattern.

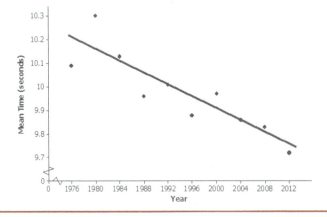

144   Lesson 11:   Using Linear Models in a Data Context

c. One reasonable line goes through the 1992 and 2004 data. Find the equation of that line.

The slope of the line through $(1992, 10.01)$ and $(2004, 9.86)$ is $\frac{10.01 - 9.86}{1992 - 2004} = -0.0125$.

To find the intercept using $(1992, 10.01)$, solve $10.01 = a + (-0.0125)(1992)$ for a, which yields $a = 34.91$.

The equation that predicts the mean medal race time for an Olympic year is $y = 34.91 + (-0.0125)x$. The mean medal race time (y) is in seconds, and the time (x) is in years.

Note to Teacher: In Algebra I, students learn a formal method called least squares for determining a "best-fitting" line. For comparison, the least squares prediction line is $y = 34.3562 + (-0.0122)x$.

d. Before he saw these data, Chang guessed that the mean time of the three Olympic medal winners decreased by about 0.05 second from one Olympic Game to the next. Does the prediction model you found in part (c) support his guess? Explain.

The slope -0.0125 means that from one calendar year to the next, the predicted mean race time for the top three medals decreases by 0.0125 second. So, between successive Olympic Games, which occur every four years, the predicted mean race time is reduced by 0.05 second because $4(0.0125) = 0.05$.

e. If the trend continues, what mean race time would you predict for the gold, silver, and bronze medal winners in the 2016 Olympic Games? Explain how you got this prediction.

If the linear pattern were to continue, the predicted mean time for the 2016 Olympics is 9.71 seconds because $34.91 - (0.0125)(2016) = 9.71$.

f. The data point $(1980, 10.3)$ appears to have an unusually high value for the mean race time (10.3). Using your library or the Internet, see if you can find a possible explanation for why that might have happened.

The mean race time in 1980 was an unusually high 10.3 seconds. In their research of the 1980 Olympic Games, students find that the United States and several other countries boycotted the games, which were held in Moscow. Perhaps the field of runners was not the typical Olympic quality as a result. Atypical points in a set of data are called outliers. They may influence the analysis of the data.

Following these two examples, ask students to summarize (in written or spoken form) how to make predictions from data.

Closing (2–3 minutes)

If time allows, revisit the linear model from Exercise 2. Explain that the data can be modified to create a model in which the y-intercept makes sense within the context of the problem.

Year	2012	2008	2004	2000	1996	1992	1988	1984	1980	1976
Number of Years (since 1976)	36	32	28	24	20	16	12	8	4	0
Gold	9.63	9.69	9.85	9.87	9.84	9.96	9.92	9.99	10.25	10.06
Silver	9.75	9.89	9.86	9.99	9.89	10.02	9.97	10.19	10.25	10.07
Bronze	9.79	9.91	9.87	10.04	9.90	10.04	9.99	10.22	10.39	10.14
Mean Time	9.72	9.83	9.86	9.97	9.88	10.01	9.96	10.13	10.30	10.09

Data Source: https://en.wikipedia.org/wiki/100_metres_at_the_Olympics#Men

Lesson 11: Using Linear Models in a Data Context

- Using the data points for 1992 and 2004, $(16, 10.01)$ and $(28, 9.86)$, the linear model is $y = 10.21 + (-0.0125)x$.
- Note that the slope is the same as the linear model in Exercise 2.
- The y-intercept is now $(0, 10.21)$, which means that in 1976 (0 years since 1976), the mean medal time was 10.21 seconds.

Review the Lesson Summary with students.

> **Lesson Summary**
>
> In the real world, it is rare that two numerical variables are exactly linearly related. If the data are roughly linearly related, then a line can be drawn through the data. This line can then be used to make predictions and to answer questions. For now, the line is informally drawn, but in later grades more formal methods for determining a best-fitting line are presented.

Exit Ticket (8–10 minutes)

Name _____   Date _____

Lesson 11: Using Linear Models in a Data Context

Exit Ticket

According to the Bureau of Vital Statistics for the New York City Department of Health and Mental Hygiene, the life expectancy at birth (in years) for New York City babies is as follows.

Year of Birth	2001	2002	2003	2004	2005	2006	2007	2008	2009
Life Expectancy	77.9	78.2	78.5	79.0	79.2	79.7	80.1	80.2	80.6

Data Source: http://www.nyc.gov/html/om/pdf/2012/pr465-12_charts.pdf

a. If you are interested in predicting life expectancy for babies born in a given year, which variable is the independent variable, and which is the dependent variable?

b. Draw a scatter plot to determine if there appears to be a linear relationship between the year of birth and life expectancy.

c. Fit a line to the data. Show your work.

d. Based on the context of the problem, interpret in words the intercept and slope of the line you found in part (c).

e. Use your line to predict life expectancy for babies born in New York City in 2010.

Exit Ticket Sample Solutions

According to the Bureau of Vital Statistics for the New York City Department of Health and Mental Hygiene, the life expectancy at birth (in years) for New York City babies is as follows.

Year of Birth	2001	2002	2003	2004	2005	2006	2007	2008	2009
Life Expectancy	77.9	78.2	78.5	79.0	79.2	79.7	80.1	80.2	80.6

Data Source: http://www.nyc.gov/html/om/pdf/2012/pr465-12_charts.pdf

a. If you are interested in predicting life expectancy for babies born in a given year, which variable is the independent variable, and which is the dependent variable?

Year of birth is the independent variable, and life expectancy in years is the dependent variable.

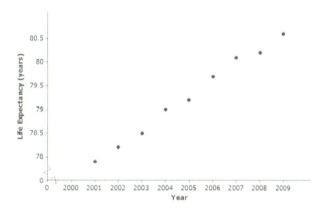

b. Draw a scatter plot to determine if there appears to be a linear relationship between the year of birth and life expectancy.

Life expectancy and year of birth appear to be linearly related.

c. Fit a line to the data. Show your work.

Answers will vary. For example, the line through $(2001, 77.9)$ and $(2009, 80.6)$ is $y = -597.438 + (0.3375)x$, where life expectancy (y) is in years, and the time (x) is in years.

Note to Teacher: The formal least squares line (Algebra I) is $y = -612.458 + (0.345)x$.

d. Based on the context of the problem, interpret in words the intercept and slope of the line you found in part (c).

Answers will vary based on part (c). The intercept says that babies born in New York City in Year 0 should expect to live around -597 years! Be sure students actually say that this is an unrealistic result and that interpreting the intercept is meaningless in this problem. Regarding the slope, for an increase of 1 in the year of birth, predicted life expectancy increases by 0.3375 year, which is a little over four months.

e. Use your line to predict life expectancy for babies born in New York City in 2010.

Answers will vary based on part (c).

$-597.438 + (0.3375)(2010) = 80.9$

Using the line calculated in part (c), the predicted life expectancy for babies born in New York City in 2010 is 80.9 years, which is also the value given on the website.

Lesson 11: Using Linear Models in a Data Context

Problem Set Sample Solutions

1. From the United States Bureau of Census website, the population sizes (in millions of people) in the United States for census years 1790–2010 are as follows.

Year	1790	1800	1810	1820	1830	1840	1850	1860	1870	1880	1890
Population Size	3.9	5.3	7.2	9.6	12.9	17.1	23.2	31.4	38.6	50.2	63.0

Year	1900	1910	1920	1930	1940	1950	1960	1970	1980	1990	2000	2010
Population Size	76.2	92.2	106.0	123.2	132.2	151.3	179.3	203.3	226.5	248.7	281.4	308.7

a. If you wanted to be able to predict population size in a given year, which variable would be the independent variable, and which would be the dependent variable?

Population size (dependent variable) is being predicted based on year (independent variable).

b. Draw a scatter plot. Does the relationship between year and population size appear to be linear?

The relationship between population size and year of birth is definitely nonlinear. Note that investigating nonlinear relationships is the topic of the next two lessons.

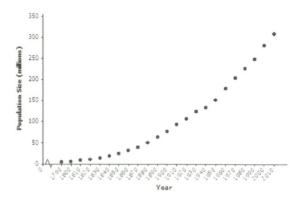

c. Consider the data only from 1950 to 2010. Does the relationship between year and population size for these years appear to be linear?

Drawing a scatter plot using the 1950–2010 data indicates that the relationship between population size and year of birth is approximately linear, although some students may say that there is a very slight curvature to the data.

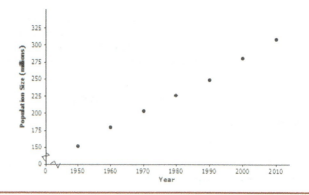

d. One line that could be used to model the relationship between year and population size for the data from 1950 to 2010 is $y = -4875.021 + 2.578x$. Suppose that a sociologist believes that there will be negative consequences if population size in the United States increases by more than $2\frac{3}{4}$ million people annually. Should she be concerned? Explain your reasoning.

This problem is asking students to interpret the slope. Some students will no doubt say that the sociologist need not be concerned, since the slope of 2.578 million births per year is smaller than her threshold value of 2.75 million births per year. Other students may say that the sociologist should be concerned, since the difference between 2.578 and 2.75 is only 172,000 births per year.

e. Assuming that the linear pattern continues, use the line given in part (d) to predict the size of the population in the United States in the next census.

The next census year is 2020.

$$-4875.021 + (2.578)(2020) = 332.539$$

The given line predicts that the population then will be 332.539 million people.

2. In search of a topic for his science class project, Bill saw an interesting YouTube video in which dropping mint candies into bottles of a soda pop caused the soda pop to spurt immediately from the bottle. He wondered if the height of the spurt was linearly related to the number of mint candies that were used. He collected data using 1, 3, 5, and 10 mint candies. Then, he used two-liter bottles of a diet soda and measured the height of the spurt in centimeters. He tried each quantity of mint candies three times. His data are in the following table.

Number of Mint Candies	1	1	1	3	3	3	5	5	5	10	10	10
Height of Spurt (centimeters)	40	35	30	110	105	90	170	160	180	400	390	420

a. Identify which variable is the independent variable and which is the dependent variable.

Height of spurt is the dependent variable, and number of mint candies is the independent variable because height of spurt is being predicted based on number of mint candies used.

Scaffolding:
- The word *spurt* may need to be defined for English language learners.
- A spurt is a sudden stream of liquid or gas forced out under pressure. Showing a visual aid to accompany this exercise may help student comprehension.

b. Draw a scatter plot that could be used to determine whether the relationship between height of spurt and number of mint candies appears to be linear.

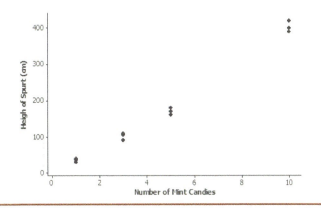

c. Bill sees a slight curvature in the scatter plot, but he thinks that the relationship between the number of mint candies and the height of the spurt appears close enough to being linear, and he proceeds to draw a line. His eyeballed line goes through the mean of the three heights for three mint candies and the mean of the three heights for 10 candies. Bill calculates the equation of his eyeballed line to be

$$y = -27.617 + (43.095)x,$$

where the height of the spurt (y) in centimeters is based on the number of mint candies (x). Do you agree with this calculation? He rounded all of his calculations to three decimal places. Show your work.

Yes, Bill's equation is correct.

The slope of the line through $(3, 101.667)$ and $(10, 403.333)$ is $\frac{403.333 - 101.667}{10 - 3} = 43.095$.

The intercept could be found by solving $403.333 = a + (43.095)(10)$ for a, which yields $a = -27.617$.

So, a possible prediction line is $y = -27.617 + (43.095)x$.

d. In the context of this problem, interpret in words the slope and intercept for Bill's line. Does interpreting the intercept make sense in this context? Explain.

The slope is 43.095, which means that for every mint candy dropped into the bottle of soda pop, the height of the spurt increases by 43.095 cm.

The y-intercept is $(0, -27.617)$. This means that if no mint candies are dropped into the bottle of soda pop, the height of the spurt is -27.617 ft. This does not make sense within the context of the problem.

e. If the linear trend continues for greater numbers of mint candies, what do you predict the height of the spurt to be if 15 mint candies are used?

$$-27.617 + (43.095)(15) = 618.808$$

The predicted height would be 618.808 cm, which is slightly over 20 ft.

Lesson 12: Nonlinear Models in a Data Context (Optional)

Student Outcomes

- Students give verbal descriptions of how y changes as x changes given the graph of a nonlinear function.
- Students draw nonlinear functions that are consistent with a verbal description of a nonlinear relationship.

Lesson Notes

This lesson is included as an optional extension to provide a deeper understanding of the key features of linear relationships in contrast to nonlinear ones.

Previous lessons focused on finding the equation of a line and interpreting the slope and intercept for data that followed a linear pattern. In the next two lessons, the focus shifts to data that do not follow a linear pattern. Instead of drawing lines through data, a curve is used to describe the relationship observed in a scatter plot.

In this lesson, students calculate the change in height of plants grown in beds with and without compost. The change in growth in the non-compost beds approximately follows a linear pattern. The change in growth in the compost beds follows a curved pattern rather than a linear pattern. Students are asked to compare the growth changes and recognize that the change in growth for a linear pattern shows a constant change, while nonlinear patterns show a rate of growth that is not constant.

Classwork

Example 1 (3 minutes): Growing Dahlias

Present the experiment for the two methods of growing dahlias. One method was to plant eight dahlias in a bed of soil that has no compost. The other was to plant eight dahlias in a bed of soil that has been enriched with compost. Explain that the students measured the height of each plant at the end of each week and recorded the median height of the eight dahlias.

Before students begin Example 1, ask the following:

- Is there a pattern in the median height of the plants?
  - *The median height is increasing every week by about* 3.5 *inches.*

Scaffolding:

- An image of a growth experiment may help English language learners understand the context of the example.
- The words *compost* and *bed* may be unfamiliar to students in this context.
- Compost is a mixture of decayed plants and other organic matter used by gardeners to enrich soil.
- *Bed* has multiple meanings. In this context, *bed* refers to a section of ground planted with flowers.
- Showing visuals of these terms to accompany the exercises aids in student comprehension.

A STORY OF RATIOS　　　　　　　　　　　　　　　　　　　　　　　Lesson 12　8•6

Example 1: Growing Dahlias

A group of students wanted to determine whether or not compost is beneficial in plant growth. The students used the dahlia flower to study the effect of composting. They planted eight dahlias in a bed with no compost and another eight plants in a bed with compost. They measured the height of each plant over a 9-week period. They found the median growth height for each group of eight plants. The table below shows the results of the experiment for the dahlias grown in non-compost beds.

Week	Median Height in Non-Compost Bed (inches)
1	9.00
2	12.75
3	16.25
4	19.50
5	23.00
6	26.75
7	30.00
8	33.75
9	37.25

Scaffolding:
Median is developed in Grades 6 and 7 as a measure of center that is used to identify a typical value for a skewed data distribution.

Exercises 1–7 (13 minutes)

This exercise set is designed as a review of the previous lesson on fitting a line to data.

The scatter plot shows that a line fits the data reasonably well. Exercise 3 asks students to find only the slope of the line. Consider having students write the equation of the line. They could then use this equation to help answer Exercise 7.

As students complete the table in Exercise 4, emphasize how the values of the change in height are all approximately equal and that they center around the value of the slope of the line that they have drawn.

Allow students to work in small groups to complete the exercises. Discuss the answers as a class.

Exercises 1–15

1. On the grid below, construct a scatter plot of non-compost height versus week.

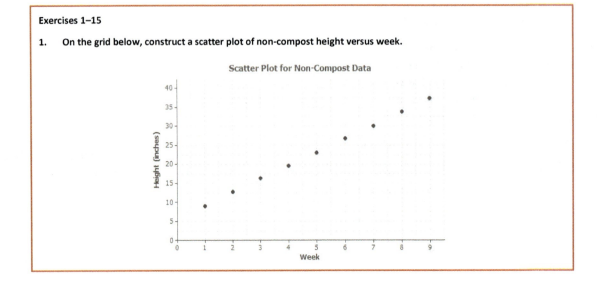

Scatter Plot for Non-Compost Data

154　　Lesson 12:　Nonlinear Models in a Data Context (Optional)

2. Draw a line that you think fits the data reasonably well.

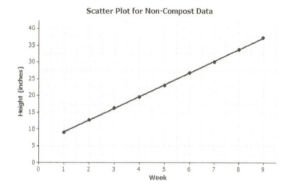

3. Find the rate of change of your line. Interpret the rate of change in terms of growth (in height) over time.

   Most students should have a rate of change of approximately 3.5 inches per week. A rate of change of 3.5 means that the median height of the eight dahlias increased by about 3.5 inches each week.

4. Describe the growth (change in height) from week to week by subtracting the previous week's height from the current height. Record the weekly growth in the third column in the table below. The median growth for the dahlias from Week 1 to Week 2 was 3.75 inches (i.e., $12.75 - 9.00 = 3.75$).

Week	Median Height in Non-Compost Bed (inches)	Weekly Growth (inches)
1	9.00	—
2	12.75	3.75
3	16.25	3.5
4	19.50	3.25
5	23.00	3.5
6	26.75	3.75
7	30.00	3.25
8	33.75	3.75
9	37.25	3.5

5. As the number of weeks increases, describe how the weekly growth is changing.

   The growth each week remains about the same—approximately 3.5 inches.

6. How does the growth each week compare to the slope of the line that you drew?

   The amount of growth per week varies from 3.25 to 3.75 but centers around 3.5, which is the slope of the line.

7. Estimate the median height of the dahlias at $8\frac{1}{2}$ weeks. Explain how you made your estimate.

   An estimate is 35.5 inches. Students can use the graph, the table, or the equation of their line.

Lesson 12: Nonlinear Models in a Data Context (Optional)

Exercises 8–14 (13 minutes)

These exercises present a set of data that do not follow a linear pattern. Students are asked to draw a curve through the data that they think fits the data reasonably well. Students may want to connect the ordered pairs, but encourage them to draw a smooth curve. A piece of thread or string can be used to sketch a smooth curve rather than connecting the ordered pairs. In this lesson, it is not expected that students find a function (nor are they given a function) that would fit the data. The main focus is that the rate of growth is not a constant when the data do not follow a linear pattern.

Allow students to work in small groups to complete the exercises. Then, discuss answers as a class.

The table below shows the results of the experiment for the dahlias grown in compost beds.

Week	Median Height in Compost Bed (inches)
1	10.00
2	13.50
3	17.75
4	21.50
5	30.50
6	40.50
7	65.00
8	80.50
9	91.50

8. Construct a scatter plot of height versus week on the grid below.

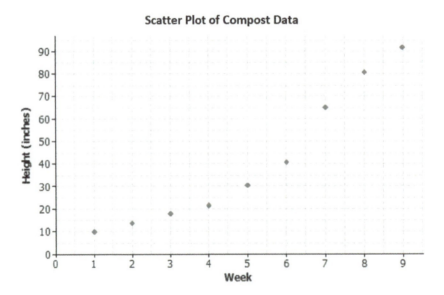

9. Do the data appear to form a linear pattern?

   No, the pattern in the scatter plot is curved.

10. Describe the growth from week to week by subtracting the height from the previous week from the current height. Record the weekly growth in the third column in the table below. The median weekly growth for the dahlias from Week 1 to Week 2 is 3.5 inches. (i.e., $13.5 - 10 = 3.5$).

Week	Compost Height (inches)	Weekly Growth (inches)
1	10.00	–
2	13.50	3.50
3	17.75	4.25
4	21.50	3.75
5	30.50	9.0
6	40.50	10.0
7	65.00	24.5
8	80.50	15.50
9	91.50	11.0

11. As the number of weeks increases, describe how the growth changes.

    The amount of growth per week varies from week to week. In Weeks 1 through 4, the growth is around 4 inches each week. From Weeks 5 to 7, the amount of growth increases, and then the growth slows down for Weeks 8 and 9.

12. Sketch a curve through the data. When sketching a curve, do not connect the ordered pairs, but draw a smooth curve that you think reasonably describes the data.

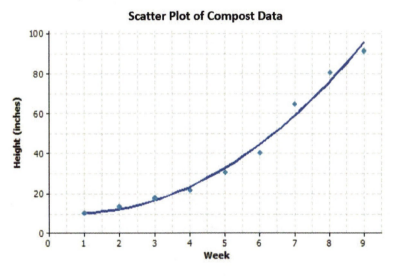

13. Use the curve to estimate the median height of the dahlias at $8\frac{1}{2}$ weeks. Explain how you made your estimate.

    Answers will vary. A reasonable estimate of the median height at $8\frac{1}{2}$ weeks is approximately 85 inches. Starting at $8\frac{1}{2}$ on the x-axis, move up to the curve and then over to the y-axis for the estimate of the height.

14. How does the weekly growth of the dahlias in the compost beds compare to the weekly growth of the dahlias in the non-compost beds?

    The growth in the non-compost is about the same each week. The growth in the compost starts the same as the non-compost, but after four weeks, the dahlias begin to grow at a faster rate.

Exercise 15 (7 minutes)

15. When there is a car accident, how do the investigators determine the speed of the cars involved? One way is to measure the skid marks left by the cars and use these lengths to estimate the speed.

    The table below shows data collected from an experiment with a test car. The first column is the length of the skid mark (in feet), and the second column is the speed of the car (in miles per hour).

Skid-Mark Length (feet)	Speed (miles per hour)
5	10
17	20
65	40
105	50
205	70
265	80

    Data Source: http://forensicdynamics.com/stopping-braking-distance-calculator
    (Note: Data has been rounded.)

    a. Construct a scatter plot of speed versus skid-mark length on the grid below.

    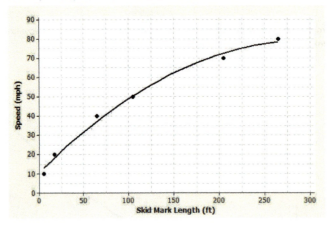

    b. The relationship between speed and skid-mark length can be described by a curve. Sketch a curve through the data that best represents the relationship between skid-mark length and the speed of the car. Remember to draw a smooth curve that does not just connect the ordered pairs.

    See the plot above.

    c. If the car left a skid mark of 60 ft., what is an estimate for the speed of the car? Explain how you determined the estimate.

    The speed is approximately 38 mph. Using the graph, for a skid mark of 65 ft., the speed was 40 mph, so the estimate is slightly less than 40 mph.

    d. A car left a skid mark of 150 ft. Use the curve you sketched to estimate the speed at which the car was traveling.

    62.5 mph

e. If a car leaves a skid mark that is twice as long as another skid mark, was the car going twice as fast? Explain.

No. When the skid mark was 105 ft. *long, the car was traveling* 50 mph. *When the skid mark was* 205 ft. *long (about twice the* 105 ft.*), the car was traveling* 70 mph, *which is not twice as fast.*

Closing (1 minute)

Review the Lesson Summary with students.

> **Lesson Summary**
>
> When data follow a linear pattern, they can be represented by a linear function whose rate of change can be used to answer questions about the data. When data do not follow a linear pattern, then there is no constant rate of change.

Exit Ticket (8 minutes)

Lesson 12: Nonlinear Models in a Data Context (Optional)

Exit Ticket

The table shows the population of New York City from 1850 to 2000 for every 50 years.

Year	Population	Population Growth (change over 50-year time period)
1850	515,547	—
1900	3,437,202	
1950	7,891,957	
2000	8,008,278	

Data Source: www.census.gov

1. Find the growth of the population from 1850 to 1900. Write your answer in the table in the row for the year 1900.

2. Find the growth of the population from 1900 to 1950. Write your answer in the table in the row for the year 1950.

3. Find the growth of the population from 1950 to 2000. Write your answer in the table in the row for the year 2000.

4. Does it appear that a linear model is a good fit for the data? Why or why not?

5. Describe how the population changes as the years increase.

6. Construct a scatter plot of time versus population on the grid below. Draw a line or curve that you feel reasonably describes the data.

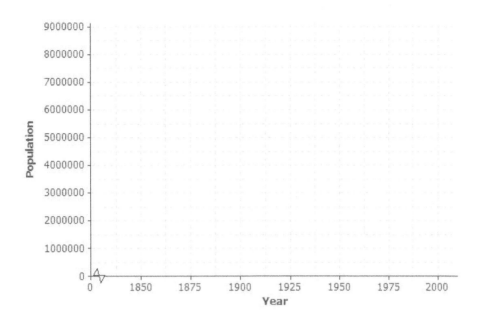

7. Estimate the population of New York City in 1975. Explain how you found your estimate.

Exit Ticket Sample Solutions

The table shows the population of New York City from 1850 to 2000 for every 50 years.

Year	Population	Population Growth (change over 50-year time period)
1850	515,547	—
1900	3,437,202	2,921,655
1950	7,891,957	4,454,755
2000	8,008,278	116,321

Data Source: www.census.gov

1. Find the growth of the population from 1850 to 1900. Write your answer in the table in the row for the year 1900.

2. Find the growth of the population from 1900 to 1950. Write your answer in the table in the row for the year 1950.

3. Find the growth of the population from 1950 to 2000. Write your answer in the table in the row for the year 2000.

4. Does it appear that a linear model is a good fit for the data? Why or why not?

   No, a linear model is not a good fit for the data. The rate of population growth is not constant; the values in the change in population column are all different.

5. Describe how the population changes as the years increase.

   As the years increase, the population increases.

6. Construct a scatter plot of time versus population on the grid below. Draw a line or curve that you feel reasonably describes the data.

   Students should sketch a curve. If students use a straight line, point out that the line does not reasonably describe the data, as some of the data points are far away from the line.

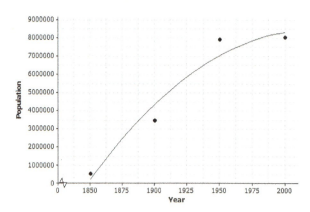

7. Estimate the population of New York City in 1975. Explain how you found your estimate.

   It is approximately 8,000,000. An estimate can be found by recognizing that the growth of the city did not change very much from 1950 to 2000. The mean of the 1950 population and the 2000 population could also be found.

Problem Set Sample Solutions

1. Once the brakes of the car have been applied, the car does not stop immediately. The distance that the car travels after the brakes have been applied is called the *braking distance*. The table below shows braking distance (how far the car travels once the brakes have been applied) and the speed of the car.

Speed (miles per hour)	Braking Distance (feet)
10	5
20	17
30	37
40	65
50	105
60	150
70	205
80	265

   Data Source: http://forensicdynamics.com/stopping-braking-distance-calculator
   (Note: Data has been rounded.)

   a. Construct a scatter plot of braking distance versus speed on the grid below.

   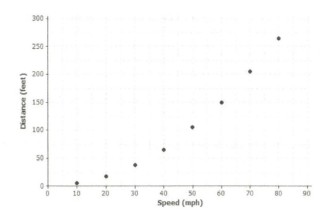

   b. Find the amount of additional distance a car would travel after braking for each speed increase of 10 mph. Record your answers in the table below.

Speed (miles per hour)	Braking Distance (feet)	Amount of Distance Increase
10	5	—
20	17	12
30	37	20
40	65	28
50	105	40
60	150	45
70	205	55
80	265	60

   c. Based on the table, do you think the data follow a linear pattern? Explain your answer.

   No. If the relationship is linear, the values in the Amount of Distance Increase column would be approximately equal.

d. Describe how the distance it takes a car to stop changes as the speed of the car increases.

As the speed of the car increases, the distance it takes the car to stop also increases.

e. Sketch a smooth curve that you think describes the relationship between braking distance and speed.

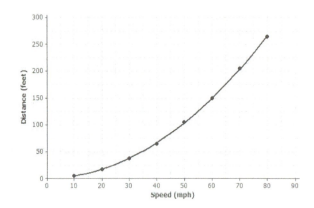

f. Estimate braking distance for a car traveling at 52 mph. Estimate braking distance for a car traveling at 75 mph. Explain how you made your estimates.

For 52 mph, the braking distance is about 115 ft.

For 75 mph, the braking distance is about 230 ft.

Both estimates can be made by starting on the x-axis, moving up to the curve, and then moving over to the y-axis.

2. The scatter plot below shows the relationship between cost (in dollars) and radius length (in meters) of fertilizing different-sized circular fields. The curve shown was drawn to describe the relationship between cost and radius.

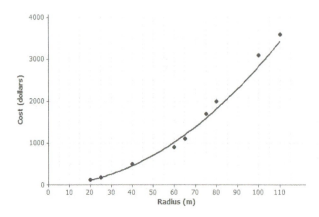

a. Is the curve a good fit for the data? Explain.

Yes, the curve fits the data very well. The data points lie close to the curve.

b. Use the curve to estimate the cost for fertilizing a circular field of radius 30 m. Explain how you made your estimate.

Using the curve drawn on the graph, the cost is approximately $200–$250.

c. Estimate the radius of the field if the fertilizing cost was $2,500. Explain how you made your estimate.

Using the curve, an estimate for the radius is approximately 94 m. Locate the approximate cost of $2,500. The approximate radius for that point is 94 m.

3. Suppose a dolphin is fitted with a GPS that monitors its position in relationship to a research ship. The table below contains the time (in seconds) after the dolphin is released from the ship and the distance (in feet) the dolphin is from the research ship.

Time (seconds)	Distance from the Ship (feet)	Increase in Distance from the Ship
0	0	—
50	85	85
100	190	105
150	398	208
200	577	179
250	853	276
300	1,122	269

a. Construct a scatter plot of distance versus time on the grid below.

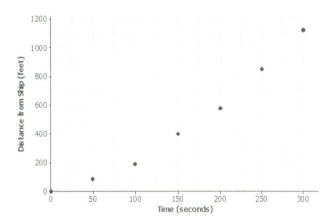

b. Find the additional distance the dolphin traveled for each increase of 50 seconds. Record your answers in the table above.

See the table above.

c. Based on the table, do you think that the data follow a linear pattern? Explain your answer.

No, the change in distance from the ship is not constant.

d. Describe how the distance that the dolphin is from the ship changes as the time increases.

As the time away from the ship increases, the distance the dolphin is from the ship is also increasing. The farther the dolphin is from the ship, the faster it is swimming.

Lesson 12: Nonlinear Models in a Data Context (Optional)

e. Sketch a smooth curve that you think fits the data reasonably well.

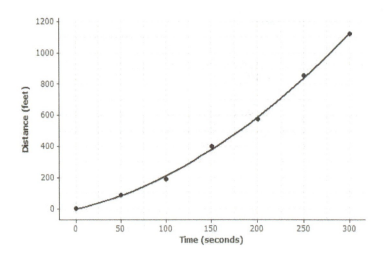

f. Estimate how far the dolphin will be from the ship after 180 seconds. Explain how you made your estimate.

About 500 ft. Starting on the x-axis at approximately 180 seconds, move up to the curve and then over to the y-axis to find an estimate of the distance.

A STORY OF RATIOS

Mathematics Curriculum

GRADE 8 • MODULE 6

Topic D
Bivariate Categorical Data

Focus Standard:	▪ Understand that patterns of association can also be seen in bivariate categorical data by displaying frequencies and relative frequencies in a two-way table. Construct and interpret a two-way table summarizing data on two categorical variables collected from the same subjects. Use relative frequencies calculated for rows or columns to describe possible association between the two variables. *For example, collect data from students in your class on whether or not they have a curfew on school nights and whether or not they have assigned chores at home. Is there evidence that those who have a curfew also tend to have chores?*
Instructional Days:	2
Lesson 13:	Summarizing Bivariate Categorical Data in a Two-Way Table (P)[1]
Lesson 14:	Association Between Categorical Variables (P)

Topic D extends the concept of a relationship between variables to bivariate categorical data. In Lesson 13, students organize bivariate categorical data into a two-way table. They calculate row and column relative frequencies and interpret them in the context of a problem. They informally decide if there is an association between two categorical variables by examining the differences of row or column relative frequencies. They interpret association between two categorical variables as knowing the value of one of the variables provides information about the likelihood of the different possible values of the other variable.

---

[1] Lesson Structure Key: **P**-Problem Set Lesson, **M**-Modeling Cycle Lesson, **E**-Exploration Lesson, **S**-Socratic Lesson

A STORY OF RATIOS

Lesson 13 8•6

 Lesson 13: Summarizing Bivariate Categorical Data in a Two-Way Table

Student Outcomes

- Students organize bivariate categorical data into a two-way table.
- Students calculate row and column relative frequencies and interpret them in context.

Lesson Notes

In this lesson, students first organize data from a survey on a single categorical variable (i.e., a univariate categorical data) into a one-way frequency table. Some questions review content on random and representative samples that students first encountered in Grade 7. Then, they organize data on two categorical variables (i.e., bivariate categorical data) into two-way frequency tables. This lesson also introduces students to relative frequencies (e.g., row and column relative frequencies). Students then interpret relative frequencies in context.

Classwork

Exercises 1–5 (3–5 minutes)

Read the opening scenario to the class. Allow students a few minutes to choose their favorite ice cream flavors. Consider also asking students to raise their hands for each flavor preference and having them record the class data in the table provided for Exercise 1.

> **Exercises 1–8**
>
> On an upcoming field day at school, the principal wants to provide ice cream during lunch. She offers three flavors: chocolate, strawberry, and vanilla. She selected your class to complete a survey to help her determine how much of each flavor to buy.
>
> 1. Answer the following question. Wait for your teacher to count how many students selected each flavor. Then, record the class totals for each flavor in the table below.
>
>   "Which of the following three ice cream flavors is your favorite: chocolate, strawberry, or vanilla?"
>
>   *Answers will vary. One possibility is shown below.*
>
> | Ice Cream Flavor | Chocolate | Strawberry | Vanilla | Total |
> |---|---|---|---|---|
> | Number of Students | 17 | 4 | 7 | 28 |
>
> 2. Which ice cream flavor do most students prefer?
>
>   *Students should respond with the most-selected flavor. For the data set shown here, that is chocolate.*
>
> 3. Which ice cream flavor do the fewest students prefer?
>
>   *Students should respond with the least-selected flavor. For the data set shown here, that is strawberry.*

168   Lesson 13:   Summarizing Bivariate Categorical Data in a Two-Way Table

A STORY OF RATIOS   Lesson 13   8•6

4. What percentage of students preferred each flavor? Round to the nearest tenth of a percent.

   Answers will vary based on the data gathered in Exercise 1.

   Chocolate: $\frac{17}{28} \approx 60.7\%$

   Strawberry: $\frac{4}{28} \approx 14.3\%$

   Vanilla: $\frac{7}{28} = 25\%$

5. Do the numbers in the table in Exercise 1 summarize data on a categorical variable or a numerical variable?

   The numbers in this table summarize data on a categorical variable—the preferred flavor of ice cream.

Scaffolding:

Categorical variables are variables that represent categorical data. Data that represent specific descriptions or categories are called categorical data.

Exercises 6–8 (5 minutes)

Let students work with a partner to discuss and answer Exercises 6–8. These exercises review the concepts of random samples and representative samples from Grade 7. These exercises may also be used to structure a class discussion.

6. Do the students in your class represent a random sample of all the students in your school? Why or why not? Discuss this with your neighbor.

   No, because there is no indication that the students were selected randomly.

7. Is your class representative of all the other classes at your school? Why or why not? Discuss this with your neighbor.

   This class might be representative of the other eighth-grade classes but might not be representative of sixth- and seventh-grade classes.

8. Do you think the principal will get an accurate estimate of the proportion of students who prefer each ice cream flavor for the whole school using only your class? Why or why not? Discuss this with your neighbor.

   It is unlikely to give a good estimate. It would depend on how representative the class is of all of the students at the school.

Example 1 (3–5 minutes)

In this example, be sure that students understand the vocabulary. *Univariate* means *one* variable. Thus, *univariate categorical data* means that there are data on one variable that are categorical, such as favorite ice cream flavor. A *one-way frequency table* is typically used to summarize values of univariate categorical data. When the data are categorical, it is customary to convert the table entries to *relative frequencies* instead of frequencies. In other words, the fraction $\frac{\text{frequency}}{\text{total}}$ should be used, which is the relative frequency or proportion for each possible value of the categorical variable.

Scaffolding:

- Point out the prefix *uni* - means one. So, *univariate* means *one variable*.
- Some students may recognize the word *table* but may not yet know the mathematical meaning of the term. Point out that this lesson defines *table* as a tool for organizing data.

Lesson 13:   Summarizing Bivariate Categorical Data in a Two-Way Table   169

A STORY OF RATIOS   Lesson 13   8•6

Students in another class were asked the same question about their favorite ice cream flavors. In this particular class of 25 students, 11 preferred chocolate, 4 preferred strawberry, and 10 preferred vanilla. Thus, the relative frequency for chocolate is $\frac{11}{25} = 0.44$. The interpretation of this value is "44% of the students in this class prefer chocolate ice cream." Students often find writing interpretations to be difficult. Explain why this is not the case in this example.

> *Scaffolding:*
>
> The word *relative* has multiple meanings, such as a family member. In this context, it refers to a measure that is compared to something else. Making this distinction clear aids in comprehension.

Example 1

Students in a different class were asked the same question about their favorite ice cream flavors. The table below shows the ice cream flavors and the number of students who chose each flavor for that particular class. This table is called a *one-way frequency table* because it shows the counts of a univariate categorical variable.

This is the univariate categorical variable. →

These are the counts for each category. →

Ice Cream Flavor	Chocolate	Strawberry	Vanilla	Total
Number of Students	11	4	10	25

We compute the relative frequency for each ice cream flavor by dividing the count by the total number of observations.

$$\text{relative frequency} = \frac{\text{count for a category}}{\text{total number of observations}}$$

Since 11 out of 25 students answered *chocolate*, the relative frequency would be $\frac{11}{25} = 0.44$. This relative frequency shows that 44% of the class prefers chocolate ice cream. In other words, the relative frequency is the proportional value that each category is of the whole.

Exercises 9–10 (3 minutes)

Let students work independently and confirm their answers with a neighbor.

Exercises 9–10

Use the table for the preferred ice cream flavors from the class in Example 1 to answer the following questions.

9. What is the relative frequency for the category *strawberry*?

    Relative frequency $= \frac{4}{25} = 0.16$

10. Write a sentence interpreting the relative frequency value in the context of strawberry ice cream preference.

    16% of the students in this class prefer strawberry ice cream.

---

170   Lesson 13:   Summarizing Bivariate Categorical Data in a Two-Way Table

©2018 Great Minds®. eureka-math.org

A STORY OF RATIOS   Lesson 13   8•6

Example 2 (3–5 minutes)

Read through the example as a class. In this example, the focus shifts to bivariate categorical data. The prefix *bi-* means two, so these data contain values for two variables that are both categorical, such as favorite ice cream flavor and gender.

> **Example 2**
>
> The principal also wondered if boys and girls have different favorite ice cream flavors. She decided to redo the survey by taking a random sample of students from the school and recording both their favorite ice cream flavors and their genders. She asked the following two questions:
>
> - "Which of the following ice cream flavors is your favorite: chocolate, strawberry, or vanilla?"
> - "What is your gender: male or female?"
>
> The results of the survey are as follows:
>
> - Of the 30 students who prefer chocolate ice cream, 22 are males.
> - Of the 25 students who prefer strawberry ice cream, 15 are females.
> - Of the 27 students who prefer vanilla ice cream, 13 are males.
>
> The values of two variables, which were ice cream flavor and gender, were recorded in this survey. Since both of the variables are categorical, the data are bivariate categorical data.

Exercises 11–17 (10 minutes)

Present Exercises 11 and 12 to the class one at a time.

> **Exercises 11–17**
>
> 11. Can we display these data in a one-way frequency table? Why or why not?
>
>   *No, a one-way frequency table is for univariate data. Here we have bivariate data, so we would need to use a two-way table.*
>
> 12. Summarize the results of the second survey of favorite ice cream flavors in the following table:
>
> | | | Favorite Ice Cream Flavor | | | |
> |---|---|---|---|---|---|
> | | | Chocolate | Strawberry | Vanilla | Total |
> | Gender | Male | 22 | 10 | 13 | 45 |
> | | Female | 8 | 15 | 14 | 37 |
> | | Total | 30 | 25 | 27 | 82 |

Lesson 13:   Summarizing Bivariate Categorical Data in a Two-Way Table

Next, remind students how to calculate relative frequencies. Give students a few minutes to calculate the approximate relative frequencies and to write them in the table. A *cell relative frequency* is a cell frequency divided by the total number of observations. Let students work independently on Exercises 13–17. Discuss and confirm the answers to Exercises 16 and 17 as a class.

13. Calculate the relative frequencies of the data in the table in Exercise 12, and write them in the following table.

		Favorite Ice Cream Flavor			
		Chocolate	Strawberry	Vanilla	Total
Gender	Male	≈ 0.27	≈ 0.12	≈ 0.16	≈ 0.55
	Female	≈ 0.10	≈ 0.18	≈ 0.17	≈ 0.45
	Total	≈ 0.37	≈ 0.30	≈ 0.33	1.0

Use the relative frequency values in the table to answer the following questions:

14. What is the proportion of the students who prefer chocolate ice cream?

    0.37

15. What is the proportion of students who are female and prefer vanilla ice cream?

    0.17

16. Write a sentence explaining the meaning of the approximate relative frequency 0.55.

    Approximately 55% of students responding to the survey are males.

17. Write a sentence explaining the meaning of the approximate relative frequency 0.10.

    Approximately 10% of students responding to the survey are females who prefer chocolate ice cream.

Example 3 (3–5 minutes)

In this example, students learn that they can also use row and column totals to calculate relative frequencies. This concept provides a foundation for future work with conditional relative frequencies in Algebra I.

Point out that students need to carefully decide which total (i.e., table total, row total, or column total) they should use.

Scaffolding:
- English language learners may need a reminder about the difference between columns and rows.
- A column refers to a vertical arrangement, and a row refers to a horizontal arrangement in the table.
- Keeping a visual aid posted that labels these parts aids in comprehension.

A STORY OF RATIOS   Lesson 13   8•6

Example 3

In the previous exercises, you used the total number of students to calculate relative frequencies. These relative frequencies were the proportion of the whole group who answered the survey a certain way. Sometimes we use row or column totals to calculate relative frequencies. We call these *row relative frequencies* or *column relative frequencies*.

Below is the two-way frequency table for your reference. To calculate "the proportion of male students who prefer chocolate ice cream," divide the 22 male students who preferred chocolate ice cream by the total of 45 male students. This proportion is $\frac{22}{45} \approx 0.49$. Notice that you used the row total to make this calculation. This is a row relative frequency.

		Favorite Ice Cream Flavor			
		Chocolate	Strawberry	Vanilla	Total
Gender	Male	22	10	13	45
	Female	8	15	14	37
	Total	30	25	27	82

Exercises 18–22 (8–10 minutes)

Discuss Exercise 18 as a class. When explaining the problem, try covering the unused part of the table with paper to focus attention on the query at hand.

Exercises 18–22

In Exercise 13, you used the total number of students to calculate relative frequencies. These relative frequencies were the proportion of the whole group who answered the survey a certain way.

18. Suppose you are interested in the proportion of male students who prefer chocolate ice cream. How is this value different from "the proportion of students who are male and prefer chocolate ice cream"? Discuss this with your neighbor.

    The proportion of students who are male and prefer chocolate ice cream is $\frac{22}{82} \approx 0.27$. This proportion uses all 82 students. The proportion of male students who prefer chocolate ice cream is $\frac{22}{45} \approx 0.49$. This proportion uses only the 45 male students as its total.

Now, allow students time to answer Exercises 19–22. Discuss student answers stressing which *total* was used in the calculation.

19. Use the table provided in Example 3 to calculate the following relative frequencies.
    a. What proportion of students who prefer vanilla ice cream are female?

    $\frac{14}{27} \approx 0.52$

    b. What proportion of male students prefer strawberry ice cream? Write a sentence explaining the meaning of this proportion in the context of this problem.

    $\frac{10}{45} \approx 0.22$ *Twenty-two percent of male students in this survey prefer strawberry ice cream.*

Lesson 13:   Summarizing Bivariate Categorical Data in a Two-Way Table

c. What proportion of female students prefer strawberry ice cream?

$$\frac{15}{37} \approx 0.41$$

d. What proportion of students who prefer strawberry ice cream are female?

$$\frac{15}{25} \approx 0.60$$

20. A student is selected at random from this school. What would you predict this student's favorite ice cream to be? Explain why you chose this flavor.

    I would predict that the student's favorite flavor is chocolate because more students chose chocolate in the survey.

21. Suppose the randomly selected student is male. What would you predict his favorite flavor of ice cream to be? Explain why you chose this flavor.

    I would predict his favorite flavor to be chocolate because more male students chose chocolate in the survey.

22. Suppose the randomly selected student is female. What would you predict her favorite flavor of ice cream to be? Explain why you chose this flavor.

    I would predict her favorite flavor to be strawberry because more female students chose strawberry in the survey.

Closing (2 minutes)

Review the Lesson Summary with students.

Lesson Summary

- Univariate categorical data are displayed in a one-way frequency table.
- Bivariate categorical data are displayed in a two-way frequency table.
- *Relative frequency* is the frequency divided by a total ( ).
- A *cell relative frequency* is a cell frequency divided by the total number of observations.
- A *row relative frequency* is a cell frequency divided by the row total.
- A *column relative frequency* is a cell frequency divided by the column total.

Exit Ticket (5 minutes)

Name _____ Date _____

Lesson 13: Summarizing Bivariate Categorical Data in a Two-Way Table

Exit Ticket

1. Explain what the term *bivariate categorical data* means.

2. Explain how to calculate relative frequency. What is another word for *relative frequency*?

3. A random group of students are polled about how they get to school. The results are summarized in the table below.

		School Transportation Survey			
		Walk	Ride Bus	Carpool	Total
Gender	Male	9	26	9	44
	Female	8	26	24	58
	Total	17	52	33	102

a. Calculate the relative frequencies for the table above. Write them as a percent in each cell of the table. Round to the nearest tenth of a percent.

b. What is the relative frequency for the Carpool category? Write a sentence interpreting this value in the context of school transportation.

c. What is the proportion of students who are female and walk to school? Write a sentence interpreting this value in the context of school transportation.

d. A student is selected at random from this school. What would you predict this student's mode of school transportation to be? Explain.

Exit Ticket Sample Solutions

1. Explain what the term *bivariate categorical data* means.

   Bivariate categorical data means that the data set comprises data on two variables that are both categorical.

2. Explain how to calculate relative frequency. What is another word for *relative frequency*?

   Relative frequency is calculated by dividing a frequency by the total number of observations. Another word for relative frequency is proportion.

3. A random group of students are polled about how they get to school. The results are summarized in the table below.

		School Transportation Survey			
		Walk	Ride Bus	Carpool	Total
Gender	Male	9 ≈ 8.8%	26 ≈ 25.5%	9 ≈ 8.8%	44 ≈ 43.1%
	Female	7 ≈ 6.9%	26 ≈ 25.5%	25 ≈ 24.5%	58 ≈ 56.9%
	Total	16 ≈ 15.7%	52 ≈ 51.0%	34 ≈ 33.3%	102 100.0%

   a. Calculate the relative frequencies for the table above. Write them as a percent in each cell of the table. Round to the nearest tenth of a percent.

   See the completed table above.

   b. What is the relative frequency for the Carpool category? Write a sentence interpreting this value in the context of school transportation.

   The relative frequency is 0.333, or 33.3%. Approximately 33.3% of the students surveyed use a carpool to get to school.

   c. What is the proportion of students who are female and walk to school? Write a sentence interpreting this value in the context of school transportation.

   The proportion is 0.069, or 6.9%. Approximately 6.9% of the students surveyed are female and walk to school.

   d. A student is selected at random from this school. What would you predict this student's mode of school transportation to be? Explain.

   I would predict the student would ride the bus because more students in the survey chose this mode of transportation.

Lesson 13: Summarizing Bivariate Categorical Data in a Two-Way Table

Problem Set Sample Solutions

Every student at Abigail Douglas Middle School is enrolled in exactly one extracurricular activity. The school counselor recorded data on extracurricular activity and gender for all 254 eighth-grade students at the school.

The counselor's findings for the 254 eighth-grade students are the following:

- Of the 80 students enrolled in band, 42 are male.
- Of the 65 students enrolled in choir, 20 are male.
- Of the 88 students enrolled in sports, 30 are female.
- Of the 21 students enrolled in art, 9 are female.

1. Complete the table below.

		Extracurricular Activities				
		Band	Choir	Sports	Art	Total
Gender	Female	38	45	30	9	122
	Male	42	20	58	12	132
	Total	80	65	88	21	254

2. Write a sentence explaining the meaning of the frequency 38 in this table.

   The frequency of 38 represents the number of eighth-grade students who are enrolled in band and are female.

Use the table provided above to calculate the following relative frequencies.

3. What proportion of students are male and enrolled in choir?

   $\frac{20}{254} \approx 0.08$

4. What proportion of students are enrolled in a musical extracurricular activity (i.e., band or choir)?

   $\frac{80 + 65}{254} \approx 0.57$

5. What proportion of male students are enrolled in sports?

   $\frac{58}{132} \approx 0.44$

6. What proportion of students enrolled in sports are male?

   $\frac{58}{88} \approx 0.66$

Pregnant women often undergo ultrasound tests to monitor their babies' health. These tests can also be used to predict the gender of the babies, but these predictions are not always accurate. Data on the gender predicted by ultrasound and the actual gender of the baby for 1,000 babies are summarized in the two-way table below.

		Predicted Gender	
		Female	Male
Actual Gender	Female	432	48
Actual Gender	Male	130	390

7. Write a sentence explaining the meaning of the frequency 130 in this table.

   The frequency of 130 represents the number of babies who were predicted to be female but were actually male (i.e., the ultrasound prediction was not correct for these babies).

Use the table provided above to calculate the following relative frequencies.

8. What is the proportion of babies who were predicted to be male but were actually female?

   $\frac{48}{1000} = 0.048$

9. What is the proportion of incorrect ultrasound gender predictions?

   $\frac{130 + 48}{1000} = 0.178$

10. For babies predicted to be female, what proportion of the predictions were correct?

    $\frac{432}{562} \approx 0.769$

11. For babies predicted to be male, what proportion of the predictions were correct?

    $\frac{390}{438} \approx 0.890$

Lesson 14: Association Between Categorical Variables

Student Outcomes

- Students use row relative frequencies or column relative frequencies to informally determine whether there is an association between two categorical variables.

Lesson Notes

In this lesson, students consider whether conclusions are reasonable based on a two-way table. Students think about what it means to have similar row relative frequencies for all rows in a table or to have similar column relative frequencies for all columns in a table. They also consider what it means to have row relative frequencies that are not similar for all rows in the table. Students study the meaning of association between two categorical variables. For example, students are asked to predict the favorite type of movie of a person whose gender is not known, and then they are asked if knowing that the person is female would change their predictions. This lesson provides a foundation for more detailed coverage of association in Algebra I.

This lesson is designed to have students work in groups of two to three. Prior to class, prepare the list of students in each group, and arrange desks or tables to allow for group work.

Classwork

Example 1 (2–3 minutes)

Let students compare the two tables. Use the following questions to lead into a discussion about association. Some students may calculate row relative frequencies to justify their answers.

- What are the variables being recorded?
  - *Smartphone use, gender, and age*
- What can you conclude about the table Smartphone Use and Gender?
  - *Answers will vary. Possible responses: 75% of those surveyed use smartphones. The percentage is the same for males and females, which is 75%.*
- What can you conclude about the table Smartphone Use and Age?
  - *Answers will vary. Possible responses: 75% of those surveyed use smartphones. However, a larger percentage of those under 40 years old use a smartphone (90%) compared to the percentage of those 40 or older (60%).*
- If you knew that someone was 20 years old, would you expect that person to use a smartphone? Explain.
  - *Yes. Possible explanation: One would expect a young person to use a smartphone based on the results in the table because 90% of people under 40 use smartphones.*

> **Scaffolding:**
> Some English language learners may need to learn the word *smartphone*. Consider providing a visual aid.

A STORY OF RATIOS   Lesson 14   8•6

> **Example 1**
>
> Suppose a random group of people are surveyed about their use of smartphones. The results of the survey are summarized in the tables below.
>
> Smartphone Use and Gender
>
> | | Use a Smartphone | Do Not Use a Smartphone | Total |
> |---|---|---|---|
> | Male | 30 | 10 | 40 |
> | Female | 45 | 15 | 60 |
> | Total | 75 | 25 | 100 |
>
> Smartphone Use and Age
>
> | | Use a Smartphone | Do Not Use a Smartphone | Total |
> |---|---|---|---|
> | Under 40 Years of Age | 45 | 5 | 50 |
> | 40 Years of Age or Older | 30 | 20 | 50 |
> | Total | 75 | 25 | 100 |

Example 2 (2 minutes)

Read the beginning of Example 2 to the class. Ask students:

- What are the variables being recorded?
    - *Movie preference and teacher or student status*

> **Example 2**
>
> Suppose a sample of 400 participants (teachers and students) was randomly selected from the middle schools and high schools in a large city. These participants responded to the following question:
>
> Which type of movie do you prefer to watch?
>
> 1. Action (*The Avengers, Man of Steel*, etc.)
> 2. Drama (*42 (The Jackie Robinson Story), The Great Gatsby*, etc.)
> 3. Science Fiction (*Star Trek Into Darkness, World War Z*, etc.)
> 4. Comedy (*Monsters University, Despicable Me 2*, etc.)
>
> Movie preference and status (teacher or student) were recorded for each participant.

Exercises 1–7 (12–15 minutes)

Have students work in small groups. Give groups one to two minutes to answer Exercise 1, and then confirm their answers as a class.

Students should read the results of the survey. Remind them that a row relative frequency is the cell frequency divided by the corresponding row total. Allow groups to answer Exercises 2–5, and then confirm answers as a class. Give groups adequate time to discuss Exercises 6 and 7, and then discuss as a class.

Lesson 14:   Association Between Categorical Variables

181

Exercises 1–7

1. Two variables were recorded. Are these variables categorical or numerical?

   Both variables are categorical.

2. The results of the survey are summarized in the table below.

	Movie Preference				
	Action	Drama	Science Fiction	Comedy	Total
Student	120	60	30	90	300
Teacher	40	20	10	30	100
Total	160	80	40	120	400

   a. What proportion of participants who are teachers prefer action movies?

   $\frac{40}{100} = 0.40$

   b. What proportion of participants who are teachers prefer drama movies?

   $\frac{20}{100} = 0.20$

   c. What proportion of participants who are teachers prefer science fiction movies?

   $\frac{10}{100} = 0.10$

   d. What proportion of participants who are teachers prefer comedy movies?

   $\frac{30}{100} = 0.30$

The answers to Exercise 2 are called *row relative frequencies*. Notice that you divided each cell frequency in the Teacher row by the total for that row. Below is a blank relative frequency table.

Table of Row Relative Frequencies

	Movie Preference			
	Action	Drama	Science Fiction	Comedy
Student	0.40	0.20	0.10	0.30
Teacher	(a) 0.40	(b) 0.20	(c) 0.10	(d) 0.30

Write your answers from Exercise 2 in the indicated cells in the table above.

3. Find the row relative frequencies for the Student row. Write your answers in the table above.
   a. What proportion of participants who are students prefer action movies?
   b. What proportion of participants who are students prefer drama movies?
   c. What proportion of participants who are students prefer science fiction movies?
   d. What proportion of participants who are students prefer comedy movies?

   See the table above.

Lesson 14: Association Between Categorical Variables

A STORY OF RATIOS — Lesson 14 — 8•6

4. Is a participant's status (i.e., teacher or student) related to what type of movie he would prefer to watch? Why or why not? Discuss this with your group.

   No. Teachers are just as likely to prefer each movie type as students are, according to the row relative frequencies.

5. What does it mean when we say that there is *no association* between two variables? Discuss this with your group.

   Answers will vary. No association means that knowing the value of one variable does not tell anything about the value of the other variable.

6. Notice that the row relative frequencies for each movie type are the same for both the Teacher and Student rows. When this happens, we say that the two variables, movie preference and status (student or teacher), are *not* associated. Another way of thinking about this is to say that knowing if a participant is a teacher (or a student) provides no information about his movie preference.

   What does it mean if row relative frequencies are not the same for all rows of a two-way table?

   It means that there is an association or a tendency between the two variables.

   > *Scaffolding:*
   >
   > For English language learners, the concept of *no association* may be difficult. However, for students working in groups, consider explicitly modeling the thinking employed in Exercise 6.

7. You can also evaluate whether two variables are associated by looking at column relative frequencies instead of row relative frequencies. A column relative frequency is a cell frequency divided by the corresponding column total.

   For example, the column relative frequency for the Student/Action cell is $\frac{120}{160} = 0.75$.

   a. Calculate the other column relative frequencies, and write them in the table below.

   Table of Column Relative Frequencies

	Movie Preference			
	Action	Drama	Science Fiction	Comedy
Student	0.75	0.75	0.75	0.75
Teacher	0.25	0.25	0.25	0.25

   b. What do you notice about the column relative frequencies for the four columns?

   The column relative frequencies are equal for all four columns.

   c. What would you conclude about association based on the column relative frequencies?

   Because the column relative frequencies are the same for all four columns, we would conclude that there is no association between movie preference and status.

In this part of the lesson, students should understand that there is a mathematical way to determine if there is no association between two categorical variables. Students can look to see if the row relative frequencies are the same (or approximately the same) for each row in the table. Discuss the mathematical method for determining if there is no association between two categorical variables.

Lesson 14: Association Between Categorical Variables

183

A STORY OF RATIOS   Lesson 14   8•6

Example 3 (2 minutes)

Introduce the data in Example 3. Give students a moment to read the results. Take a quick movie preference poll in class. Ask the following:

- Who likes action movies?
- Do you think movie preference is equal among males and females?
    - *Answers will vary. Encourage students to explain why they think the preferences might be equal or different.*

Example 3

In the survey described in Example 2, gender for each of the 400 participants was also recorded. Some results of the survey are given below:

- 160 participants preferred action movies.
- 80 participants preferred drama movies.
- 40 participants preferred science fiction movies.
- 240 participants were females.
- 78 female participants preferred drama movies.
- 32 male participants preferred science fiction movies.
- 60 female participants preferred action movies.

Exercises 8–11 (8–10 minutes)

Let students work with their groups on Exercises 8–10, and then confirm answers as a class. Give students two to three minutes to complete Exercise 11.

Exercises 8–15

Use the results from Example 3 to answer the following questions. Be sure to discuss these questions with your group members.

8. Complete the two-way frequency table that summarizes the data on movie preference and gender.

	Movie Preference				
	Action	Drama	Science Fiction	Comedy	Total
Female	60	78	8	94	240
Male	100	2	32	26	160
Total	160	80	40	120	400

9. What proportion of the participants are female?

$$\frac{240}{400} = 0.60$$

10. If there was no association between gender and movie preference, should you expect more females than males or fewer females than males to prefer action movies? Explain.

If there was no association between gender and movie preference, then I would expect more females than males to prefer action movies just because there are more females in the sample. However, if there was an association between gender and movie preference, then I would expect either fewer females than males who prefer action movies or considerably more females than males who prefer action movies.

Lesson 14:   Association Between Categorical Variables

11. Make a table of row relative frequencies of each movie type for the Male row and the Female row. Refer to Exercises 2–4 to review how to complete the table below.

	Movie Preference			
	Action	Drama	Science Fiction	Comedy
Female	0.25	0.325	0.033	0.392
Male	0.625	0.0125	0.2	0.1625

Exercises 12–15 (12–15 minutes)

Read the next instructions. Make sure that students understand that 1 of the 400 participants is randomly selected. Allow groups about five minutes to discuss and answer Exercises 12 and 13.

Then, discuss as a class what *association* means. Allow students three minutes to answer Exercise 14.

Allow five minutes for groups to discuss whether the statements in Exercise 15 are correct. Call on groups to share their answers.

Suppose that you randomly pick 1 of the 400 participants. Use the table of row relative frequencies on the previous page to answer the following questions.

12. If you had to predict what type of movie this person chose, what would you predict? Explain why you made this choice.

    The participant likely prefers action movies because the largest proportion of participants preferred action movies.

13. If you know that the randomly selected participant is female, would you predict that her favorite type of movie is action? If not, what would you predict, and why?

    No. A female participant is more likely to prefer comedy since it has the greatest row relative frequency in the Female row.

14. If knowing the value of one of the variables provides information about the value of the other variable, then there is an association between the two variables.

    Is there an association between the variables gender and movie preference? Explain.

    Yes. The row relative frequencies are not the same (not even close) in each row in the table.

15. What can be said when two variables are associated? Read the following sentences. Decide if each sentence is a correct statement based upon the survey data. If it is not correct, explain why not.

    a. More females than males participated in the survey.

       Correct

    b. Males tend to prefer action and science fiction movies.

       Correct

    c. Being female causes one to prefer drama movies.

       Incorrect Association does not imply a cause-and-effect relationship.

Lesson 14: Association Between Categorical Variables

185

A STORY OF RATIOS   Lesson 14   8•6

Closing (3 minutes)

Read through the Lesson Summary with students.

If time allows, have students refer back to Example 1 and calculate row relative frequencies for each table to determine if there is evidence of association between variables.

Lesson Summary

- Saying that two variables *are not* associated means that knowing the value of one variable provides no information about the value of the other variable.

- Saying that two variables *are* associated means that knowing the value of one variable provides information about the value of the other variable.

- To determine if two variables are associated, calculate row relative frequencies. If the row relative frequencies are about the same for all of the rows, it is reasonable to say that there is no association between the two variables that define the table.

- Another way to decide if there is an association between two categorical variables is to calculate column relative frequencies. If the column relative frequencies are about the same for all of the columns, it is reasonable to say that there is no association between the two variables that define the table.

- If the row relative frequencies are quite different for some of the rows, it is reasonable to say that there is an association between the two variables that define the table.

Exit Ticket (5 minutes)

Name _____   Date _____

Lesson 14: Association Between Categorical Variables

Exit Ticket

A random sample of 100 eighth-grade students are asked to record two variables: whether they have a television in their bedrooms and if they passed or failed their last math test. The results of the survey are summarized below.

- 55 students have a television in their bedrooms.
- 35 students do not have a television in their bedrooms and passed their last math test.
- 25 students have a television and failed their last math test.
- 35 students failed their last math test.

1. Complete the two-way table.

	Pass	Fail	Total
Television in the Bedroom			
No Television in the Bedroom			
Total			

2. Calculate the row relative frequencies, and enter the values in the table above. Round to the nearest thousandth.

3. Is there evidence of association between the variables? If so, does this imply there is a cause-and-effect relationship? Explain.

Exit Ticket Sample Solutions

A random sample of 100 eighth-grade students are asked to record two variables: whether they have a television in their bedrooms and if they passed or failed their last math test. The results of the survey are summarized below.

- 55 students have a television in their bedrooms.
- 35 students do not have a television in their bedrooms and passed their last math test.
- 25 students have a television and failed their last math test.
- 35 students failed their last math test.

1. Complete the two-way table.

	Pass	Fail	Total
Television in the Bedroom	30 ≈ 0.545	25 ≈ 0.455	55 = 1.000
No Television in the Bedroom	35 ≈ 0.778	10 ≈ 0.222	45 = 1.000
Total	65 = 0.650	35 = 0.350	100 = 1.000

2. Calculate the row relative frequencies, and enter the values in the table above. Round to the nearest thousandth.

   The row relative frequencies are displayed in the table above.

3. Is there evidence of association between the variables? If so, does this imply there is a cause-and-effect relationship? Explain.

   Yes, there is evidence of association between the variables because the relative frequencies are different among the rows. However, this does not necessarily imply a cause-and-effect relationship. The fact that a student has a television in his room does not cause the student to fail a test. Rather, it may be that the student is spending more time watching television or playing video games instead of studying.

Problem Set Sample Solutions

A sample of 200 middle school students was randomly selected from the middle schools in a large city. Answers to several survey questions were recorded for each student. The tables below summarize the results of the survey.

For each table, calculate the row relative frequencies for the Female row and for the Male row. Write the row relative frequencies beside the corresponding frequencies in each table below.

1. This table summarizes the results of the survey data for the two variables, gender and which sport the students prefer to play. Is there an association between gender and which sport the students prefer to play? Explain.

		Sport				
		Football	Basketball	Volleyball	Soccer	Total
Gender	Female	2 ≈ 0.021	29 ≈ 0.299	28 ≈ 0.289	38 ≈ 0.392	97
	Male	35 ≈ 0.340	36 ≈ 0.350	8 ≈ 0.078	24 ≈ 0.233	103
	Total	37	65	36	62	200

Yes, there appears to be an association between gender and sports preference. The row relative frequencies are not the same for the Male and the Female rows, as shown in the table above.

2. This table summarizes the results of the survey data for the two variables, gender and the students' T-shirt sizes. Is there an association between gender and T-shirt size? Explain.

		School T-Shirt Sizes				
		Small	Medium	Large	X-Large	Total
Gender	Female	47 ≈ 0.484	35 ≈ 0.361	13 ≈ 0.134	2 ≈ 0.021	97
	Male	11 ≈ 0.107	41 ≈ 0.398	42 ≈ 0.408	9 ≈ 0.087	103
	Total	58	76	55	11	200

Yes, there appears to be an association between gender and T-shirt size. The row relative frequencies are not the same for the Male and the Female rows, as shown in the table above.

3. This table summarizes the results of the survey data for the two variables, gender and favorite type of music. Is there an association between gender and favorite type of music? Explain.

		Favorite Type of Music				
		Pop	Hip-Hop	Alternative	Country	Total
Gender	Female	35 ≈ 0.361	28 ≈ 0.289	11 ≈ 0.113	23 ≈ 0.237	97
	Male	37 ≈ 0.359	30 ≈ 0.291	13 ≈ 0.126	23 ≈ 0.223	103
	Total	72	58	24	46	200

No, there does not appear to be an association between gender and favorite type of music. The row relative frequencies are about the same for the Male and Female rows, as shown in the table above.

Lesson 14: Association Between Categorical Variables

Name _____ Date _____

1. The Kentucky Derby is a horse race held each year. The following scatter plot shows the speed of the winning horse at the Kentucky Derby each year between 1875 and 2012.

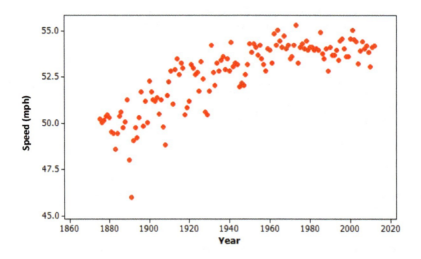

Data Source: http://www.kentuckyderby.com/

(Note: Speeds were calculated based on times given on website.)

   a. Is the association between *speed* and *year* positive or negative? Give a possible explanation in the context of this problem for why the association behaves this way considering the variables involved.

   b. Comment on whether the association between *speed* and *year* is approximately linear, and then explain in the context of this problem why the form of the association (linear or not) makes sense considering the variables involved.

c. Circle an outlier in this scatter plot, and explain, in context, how and why the observation is unusual.

2. Students were asked to report their gender and how many times a day they typically wash their hands. Of the 738 males, 66 said they wash their hands at most once a day, 583 said two to seven times per day, and 89 said eight or more times per day. Of the 204 females, 2 said they wash their hands at most once a day, 160 said two to seven times per day, and 42 said eight or more times per day.

   a. Summarize these data in a two-way table with rows corresponding to the three different frequency-of-hand-washing categories and columns corresponding to gender.

   b. Do these data suggest an association between *gender* and *frequency of hand washing*? Support your answer with appropriate calculations.

3. Basketball players who score a lot of points also tend to be strong in other areas of the game such as number of rebounds, number of blocks, number of steals, and number of assists. Below are scatter plots and linear models for professional NBA (National Basketball Association) players last season.

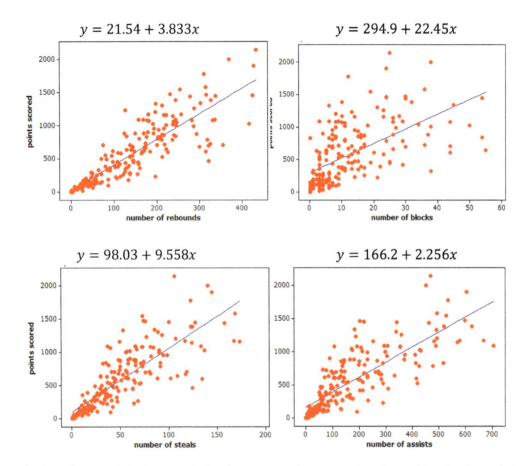

a. The line that models the association between points scored and number of rebounds is $y = 21.54 + 3.833x$, where y represents the number of points scored and x represents the number of rebounds. Give an interpretation, in context, of the slope of this line.

b. The equations on the previous page all show the number of points scored (y) as a function of the other variables. An increase in which of the variables (rebounds, blocks, steals, and assists) tends to have the largest impact on the predicted points scored by an NBA player?

c. Which of the four linear models shown in the scatter plots on the previous page has the worst fit to the data? Explain how you know using the data.

A STORY OF RATIOS   End-of-Module Assessment Task   8•6

A Progression Toward Mastery					
Assessment Task Item		STEP 1 Missing or incorrect answer and little evidence of reasoning or application of mathematics to solve the problem	STEP 2 Missing or incorrect answer but evidence of some reasoning or application of mathematics to solve the problem	STEP 3 A correct answer with some evidence of reasoning or application of mathematics to solve the problem, OR an incorrect answer with substantial evidence of solid reasoning or application of mathematics to solve the problem	STEP 4 A correct answer supported by substantial evidence of solid reasoning or application of mathematics to solve the problem
1	a	Student does not use the data in the scatter plot or context to answer the question.	Student discusses horses getting faster with newer training methods but does not discuss the data in the scatter plot.	Student discusses the overall increase of speeds but does not discuss how those data imply horses getting faster over time.	Student discusses the overall increase of speeds and how those data imply horses getting faster over time.
	b	Student does not use the data in the scatter plot or context to answer the question.	Student does not recognize the nonlinear nature of the data.	Student discusses the nonlinear nature of the data but does not relate to the context.	Student discusses the curvature in the data, which indicates that speeds should level off.
	c	Student does not use the data in the scatter plot or context to answer the question.	Student picks the year with the fastest or lowest speed of the winning horse and does not explain the choice.	Student picks the year with the lowest speed of the winning horse but does not interpret the negative residual.	Student picks the year with the lowest speed of the winning horse and states that the speed is much lower than expected for that year.
2	a	Student does not use the data given in the stem.	Student gives the tallies of the two distributions separately without looking at the cross-tabulation.	Student constructs the table but uses gender as the row variable.	Student constructs a 3×2 two-way table, including appropriate labels.

Module 6:   Linear Functions

	b	Student answer is based only on context without references to the data.	Student gives some information about the association but does not back it up numerically. OR Student says the results cannot be compared because the numbers of males and females are not equal.	Student attempts to calculate the six conditional proportions but compares them inappropriately. OR Student does not correctly complete all the calculations.	Student calculates the six conditional proportions, compares them, and draws an appropriate comparison (e.g., 20% of females wash eight or more times compared to 12% of males).
3	a	Student cannot identify the slope from the stem.	Student interprets the slope incorrectly.	Student interprets the slope correctly but not in context or not in terms of the model estimation.	Student interprets the slope correctly and predicts that on average, for each additional rebound, an increase of 3.833 points is scored.
	b	Student does not relate to the functions provided above the scatter plots.	Student focuses on the strength of the association in terms of how close the dots fall to the regression line.	Student appears to relate the question to the slope of the equation but cannot make a clear choice of which variable has the largest impact or does not provide a complete justification.	Student relates the question to the slope and identifies the number of blocks as the variable with the largest impact.
	c	Student does not use the data in the scatter plots to answer the question.	Student focuses only on the slope of the line or on one or two values that are not well predicted.	Student focuses on the vertical distances of the dots from the line but is not able to make a clear choice.	Student selects the number of blocks based on the additional spread of the dots about the regression line in that scatter plot compared to the other variables.

Module 6: Linear Functions

A STORY OF RATIOS End-of-Module Assessment Task 8•6

Name _____ Date _____

1. The Kentucky Derby is a horse race held each year. The following scatter plot shows the speed of the winning horse at the Kentucky Derby each year between 1875 and 2012.

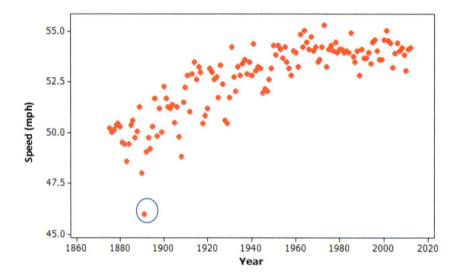

Data Source: http://www.kentuckyderby.com/
(Note: Speeds were calculated based on times given on website.)

a. Is the association between *speed* and *year* positive or negative? Give a possible explanation in the context of this problem for why the association behaves this way considering the variables involved.

> The association is positive overall, as horses have been getting faster over time. This is perhaps due to improved training methods.

b. Comment on whether the association between *speed* and *year* is approximately linear, and then explain in the context of this problem why the form of the association (linear or not) makes sense considering the variables involved.

> The association is not linear. There is probably a physical limit to how fast horses can go that we are approaching.

c. Circle an outlier in this scatter plot, and explain, in context, how and why the observation is unusual.

The winner that year was much slower than we would have predicted.

2. Students were asked to report their gender and how many times a day they typically wash their hands. Of the 738 males, 66 said they wash their hands at most once a day, 583 said two to seven times per day, and 89 said eight or more times per day. Of the 204 females, 2 said they wash their hands at most once a day, 160 said two to seven times per day, and 42 said eight or more times per day.

   a. Summarize these data in a two-way table with rows corresponding to the three different frequency-of-hand-washing categories and columns corresponding to gender.

	males	females
≤ 1	66	2
2-7	583	160
≥ 8	89	42
	738	204

   b. Do these data suggest an association between *gender* and *frequency of hand washing*? Support your answer with appropriate calculations.

	males	females
≤ 1	.0894	.0098
2-7	.7900	.7843
≥ 8	.1206	.2059

Males are more likely than females to wash hands at most once per day. Females are more likely to wash 8 or more times per day.

3. Basketball players who score a lot of points also tend to be strong in other areas of the game such as number of rebounds, number of blocks, number of steals, and number of assists. Below are scatter plots and linear models for professional NBA (National Basketball Association) players last season.

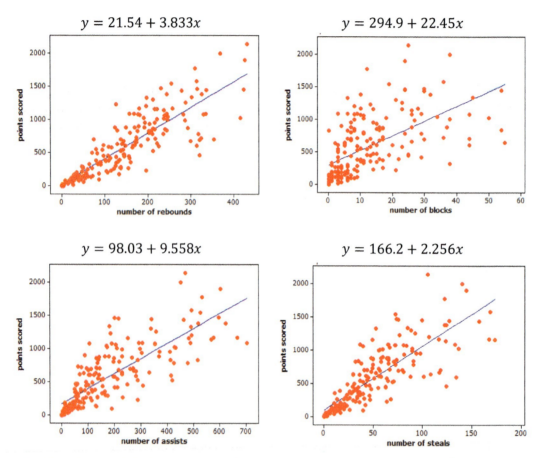

a. The line that models the association between points scored and number of rebounds is $y = 21.54 + 3.833x$, where y represents the number of points scored and x represents the number of rebounds. Give an interpretation, in context, of the slope of this line.

If the number of rebounds increases by one, we predict the number of points increases by 3.833.

b. The equations on the previous page all show the number of points scored (y) as a function of the other variables. An increase in which of the variables (rebounds, blocks, steals, and assists) tends to have the largest impact on the predicted points scored by an NBA player?

Each additional block corresponds to 22.45 more points, the largest slope or rate of increase.

c. Which of the four linear models shown in the scatter plots on the previous page has the worst fit to the data? Explain how you know using the data

Probably number of blocks because the association is weaker. There is more scatter of the points away from the line.

This page intentionally left blank

Credits

Great Minds® has made every effort to obtain permission for the reprinting of all copyrighted material. If any owner of copyrighted material is not acknowledged herein, please contact Great Minds for proper acknowledgment in all future editions and reprints of this module.

- All material from the *Common Core State Standards for Mathematics* © Copyright 2010 National Governors Association Center for Best Practices and Council of Chief State School Officers. All rights reserved.

This page intentionally left blank